DISTRIBUTION MODELS THEORY

DISTRIBUTION MODELS THEORY

editors

Rafael Herrerías Pleguezuelo

José Callejón Céspedes

José Manuel Herrerías Velasco

University of Granada, Spain

World Scientific

NEW JERSEY • LONDON • SINGAPORE • BEIJING • SHANGHAI • HONG KONG • TAIPEI • CHENNAI

Published by

World Scientific Publishing Co. Pte. Ltd.

5 Toh Tuck Link, Singapore 596224

USA office: 27 Warren Street, Suite 401-402, Hackensack, NJ 07601

UK office: 57 Shelton Street, Covent Garden, London WC2H 9HE

Library of Congress Cataloging-in-Publication Data

Distribution models theory / edited by Rafael Herrerías-Pleguezuelo,
José Callejón-Céspedes, and José Manuel Herrerías-Velasco.
p. cm.
Includes bibliographical references and index.
ISBN 981-256-900-6 (alk. paper)
1. Model theory. 2. Distribution (Probability theory). I. Herrerías-Pleguezuelo, Rafael.
II. Callejón-Céspedes, José III. Herrerías-Velasco, José Manuel.

QA9.7.D58 2006
511.3'4--dc22 2006048221

British Library Cataloguing-in-Publication Data

A catalogue record for this book is available from the British Library.

Printed in Singapore by World Scientific Printers (S) Pte Ltd

Preface

The monograph contains a compilation of papers previously chosen by the Scientific Committee of the Fifth Workshop of Spanish Scientific Association of Applied Economy on Distribution Models Theory held in Granada (Spain) in September 2005.

As editors, we endeavored to give a high scientific level in this volume. All the papers have been carefully selected, revised and presented at a high level. Therefore, this volume offers a compulsory point of reference on models theory for statisticians, economists, mathematics and in general for all researchers who are working on models theory and are eager to know the most recent advances from methodological and practical points of view.

Among the authors we appreciate the efforts of Prof. van Dorp who has made possible for us to include with pleasure his paper coauthored with Prof. Samuel Kotz, Editor-in-Chief of the *Encyclopedia of Statistical Sciences*. We would also like to acknowledge warmly all those who, through their papers, have contributed to make this high-quality volume possible. To all of them, thank you very much.

Rafael Herrerías Pleguezuelo
José Callejón Céspedes
José Manuel Herrerías Velasco
University of Granada, Spain
April 2006

Contents

Chapter 1

MODELING INCOME DISTRIBUTIONS USING ELEVATED DISTRIBUTIONS ON A BOUNDED DOMAIN

J. RENÉ VAN DORP

Engineering Management and Systems Engineering Department
The George Washington University
1776 G street, Suite 110, NW, Washington DC, 20052

SAMUEL KOTZ

Engineering Management and Systems Engineering Department
The George Washington University
1776 G street, Suite 110, NW, Washington DC, 20052

This paper presents a new two parameter family of continuous distribution on a bounded domain which has an elevated but finite density value at its lower bound. Such a characteristic appears to be useful, for example, when representing income distributions at lower income ranges. The family generalizes the one parameter Topp and Leone distribution originated in the 1950's and recently rediscovered. The family of beta distributions has been used for modeling bounded income distribution phenomena, but it only allows for an infinite and zero density values at its lower bound, and a constant density of 1 in case of its uniform member. The proposed family alleviates this apparent jump discontinuity at the lower bound. The U.S. Income distribution data for the year 2001 is used to fit distributions for Caucasian (Non-Hispanic), Hispanic and African-American populations via a maximum likelihood procedure. The results reveal stochastic ordering when comparing the Caucasian (Non-Hispanic) income distribution to that of the Hispanic or African-American population. The latter indicates that although substantial advances have reportedly been made in reducing the income distribution gap amongst different ethnic groups in the U.S. during the last 20 years or so, these differences still exist.

1. Introduction

In a 1955 issue of the Journal of the American Statistical Association an isolated paper on a bounded continuous distribution by Topp and Leone [1] appeared which received little attention. The paper was re-discovered by Nadarajah and Kotz [2] and motivated by investigations of van Dorp and Kotz [3,4] on the Two-Sided Power (TSP) distribution and other alternatives to the popular and versatile beta distribution which has been used in various applications for over a century. Even in the late nineties of the 20th century the arsenal of bounded univariate distributions contained very few members. Amongst them, the triangular and uniform distribution are the most widely used together with some "curious" distributions appearing as problems or exercises in various Mathematical and Statistical journals. Other, somewhat artificial empirical bounded continuous distributions are based on mathematical transformations of the normal distribution (of an unbounded domain) - the most wide spread amongst them are perhaps the Johnson [5] family of transformations. On the other hand the existence of multitudes of unbounded continuous distributions developed in the 20th century is well known and amply documented.

The construction of the Topp and Leone distribution is quite straightforward and based on the principle that by raising an arbitrary cdf $F(x) \in [0,1]$ to an arbitrary power $\beta > 0$, a new cdf $G(x) = F^{\beta}(x)$ emerges with one additional parameter. This devise was used in 1939 by W. Weibull [6] proposing his Weibull distribution, which has achieved substantial popularity the second part of the 20th century, especially in reliability and biometrical applications. The cdf F(x) in the above construction method may be referred to as the generating cdf. Figure 1 demonstrates the construction of the Topp and Leone distribution. The generating density of the Topp and Leone family is the right triangular density $(2-2x)$, $x \in [0,1]$. It is displayed in Figure 1A. Figure 1B depicts its cdf $(2x - 2x^2)$ and Figures 2C and 2D plot the pdf and cdf of a one parameter Topp and Leone distribution for $\beta = 3$. Note, the appearance of a mode in the

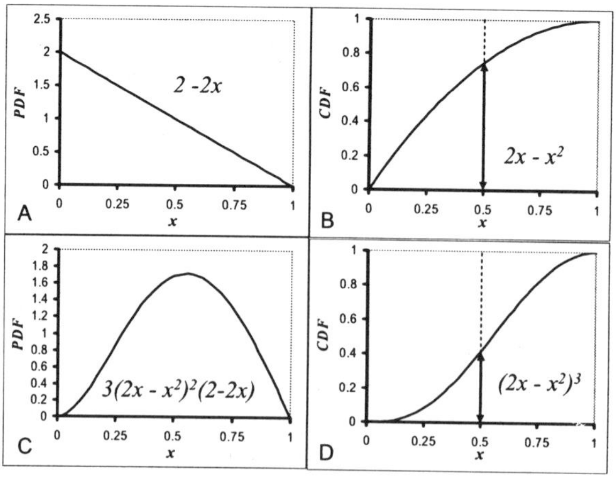

Figure 1. Construction of Topp and Leone distribution from a right triangular distribution

pdf presented in Figure 1C due to S-shapedness of the corresponding cdf in Figure 1D obtained by using a cdf transformation with $\beta > 1$. Topp and Leone's [1] original interest focused on the construction of J-shaped distributions utilizing similar cdf transformations with $0 < \beta < 1$; They have fitted their distribution to transmitter tubes failure data. Nadarajah and Kotz [2] showed that the J-shaped Topp and Leone distributions exhibit a bath tub failure rate functions with natural applications in reliability.

Our generalization of the Topp and Leone distribution (GTL) utilizes a slightly more general slope distribution with pdf $\alpha x - 2(\alpha - 1)x, 0 \le \alpha \le 2$, as the generating density (see Figure 2A with $\alpha = 1.5$), where $x \in (0,1)$. Slope distributions possess linear pdf's and play a central role in deriving a generalization of the trapezoidal distribution (see, e.g., Van Dorp and Kotz [7]). From the restriction that $\alpha x - 2(\alpha - 1)x \ge 0$ for all $x \in (0,1)$, it follows that $0 \le \alpha \le 2$. For $\alpha \in [0,1)$ $(\alpha \in (1,2])$, the slope of the pdf is increasing (decreasing). For $\alpha = 1$, the slope distribution (1) simplifies to a uniform distribution on $(0,1)$. Figure 2B plots the cdf of the linear pdf in Figure 2A.

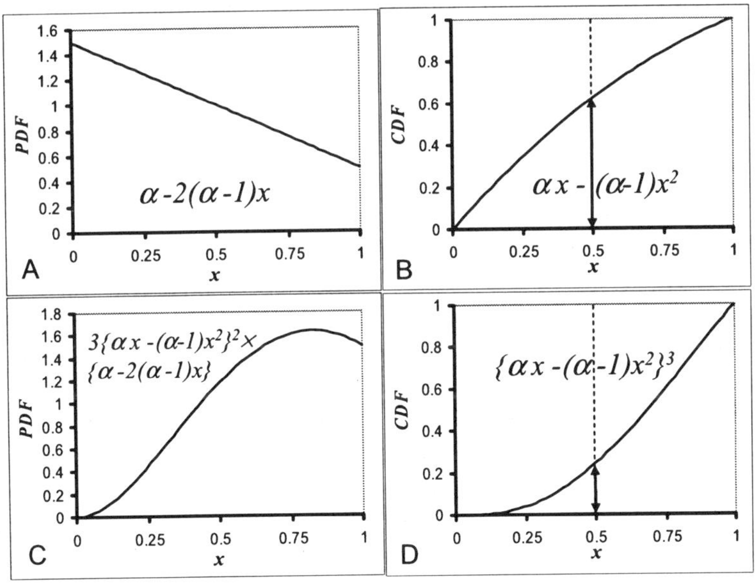

Figure 2. Construction of generalized Topp and Leone distribution from a slope distribution

Now the Generalized Topp and Leone (GTL) distribution that follows from Figure 2B (utilizing the above construction method with $\beta = 3$) is depicted in Figure 2D. The density associated with this cdf is displayed in Figure 2C. Note that, while a mode in $(0,1)$ is present in Figure 2C, it has been shifted to the right when compared to the situation in Figure 1C. More importantly, the density at the upper bound is strictly positive in Figure 2C while being zero in Figure 1C (representing the original Topp and Leone density).

Our main interest in this paper is to represent income distributions. We shall therefore consider the reflected version of the Generalized Topp and Leone (GTL) distribution utilizing the cdf transformation $H(x)=1-G(1-x)$, where G is a GTL cdf on $[0,1]$. The latter transformation typically assigns the mode towards the left hand side of its support and allows for strictly positive density values at the lower bound. This form seems to be appropriate when representing income distributions at lower income ranges. (Compare, e.g., with Figure 2 of Barsky et al. [8], p. 668). The U.S. Income distribution data for the year 2001 is used to fit Reflected GTL (RGTL) distributions for Caucasian (Non-Hispanic), Hispanic and African-American populations via a maximum likelihood procedure. The results reveal stochastic ordering when comparing the Caucasian (Non-Hispanic) income distribution to that of the Hispanic or African-American populations. In particular when comparing Americans of Caucasian Origin, African-Americans appear to be approximately 1.9 times as likely and the Hispanics 1.5 times as likely to have inadequate or no income at all. The latter indicates that although substantial advances have indeed occurred in reducing the income distribution gap amongst different ethnic groups in the U.S. during the last 20 years or so (see, e.g., Couch and Daly [9]), these differences still exist.

Another reason to consider reflected GTL distribution rather than GTL distributions is that a drift of the mode towards the left hand side mimics the behavior of the classical unbounded continuous distributions such as the Gamma, Weibull and Lognormal. (We note, in passing, that these three distributions are in a strong competition amongst themselves as to which is the best one for fitting numerous phenomena in economics, engineering and medical applications). One can therefore conjecture that application of Reflected GTL (RGTL) distributions may not be limited to the area of income distributions.

In Section 2, we shall present the cdf and pdf of a four parameter RGTL distribution and investigate its various forms. In Section 3, we will elaborate on some properties of RGTL distributions. Moment expressions for RGTL distributions, to the best of our knowledge, cannot be derived in closed form (except for certain special cases). The cdf of the beta distribution while not available in a closed form (whereas that of an RGTL distribution is) is, however, useful for calculating moments of RGTL distributions for $1<\alpha\leq 2$. In Section 4, we shall discuss a Maximum Likelihood Estimation (MLE) procedure utilizing standard root finding algorithms that are readily available in various software packages such as e.g. Microsoft Excel. In Section 5, we shall fit RGTL distributions to the U.S. 2001 income distribution data with seemingly satisfactory results. Some brief concluding remarks are presented in Section 6.

2. Cumulative distribution function and density function

The four parameter RGTL distribution with support $[a, b]$ the cdf

$$F(x|a,b,\alpha,\beta) = 1 - \left(\frac{b-x}{b-a}\right)^{\beta}\left\{\alpha-(\alpha-1)\left(\frac{b-x}{b-a}\right)\right\}^{\beta} \tag{1}$$

where $a \le x \le b$, $0 < \alpha \le 2$ and $\beta > 0$ Evidently, $F(a) = 0$ and $F(b) = 1$. The probability density function (pdf) follows from (1) to be

$$f(x|a,b,\alpha,\beta) = \frac{\beta}{b-a}\left(\frac{b-x}{b-a}\right)^{\beta-1} \times \left\{\alpha-(\alpha-1)\left(\frac{b-x}{b-a}\right)\right\}^{\beta-1}\left\{\alpha-2(\alpha-1)\left(\frac{b-x}{b-a}\right)\right\} \tag{2}$$

with the same constraint on x, α and β as in (1). From (2) it follows that in particular

$$f(a|a,b,\alpha,\beta) = \frac{\beta(2-\alpha)}{b-a} \tag{3}$$

and

$$f(b|a,b,\alpha,\beta) = \begin{cases} 0 & \beta > 1 \\ \dfrac{\beta\alpha}{b-a} & \beta = 1 \\ \to \infty \text{ as } x \uparrow b & \beta < 1 \end{cases} \tag{4}$$

Relation (3) shows that the RGTL family allows for arbitrary density values at its lower bound a. Expressions (1) and (2) are reduced to the Topp and Leone distribution (see Topp and Leone [1]):

$$f(x|a,b,\alpha,\beta) = \frac{2\beta}{b}\left(\frac{x}{b}\right)^{\beta-1}\left\{2-\left(\frac{x}{b}\right)\right\}^{\beta-1}\left\{1-\left(\frac{x}{b}\right)\right\} \tag{5}$$

by setting $a = 0$, $b = 2$ and utilizing the reflection transformation $Y = b + a - X$. Figure 1C depicts a graph of the Topp and Leone distribution with parameters $b = 1$ and $\beta = 3$ in (5). Figure 3A displays its reflected version. Note the transition in the form of graphical representations of the pdf's from Figure 3B to Figure 3D which all have the same value of α with decreasing values of β.

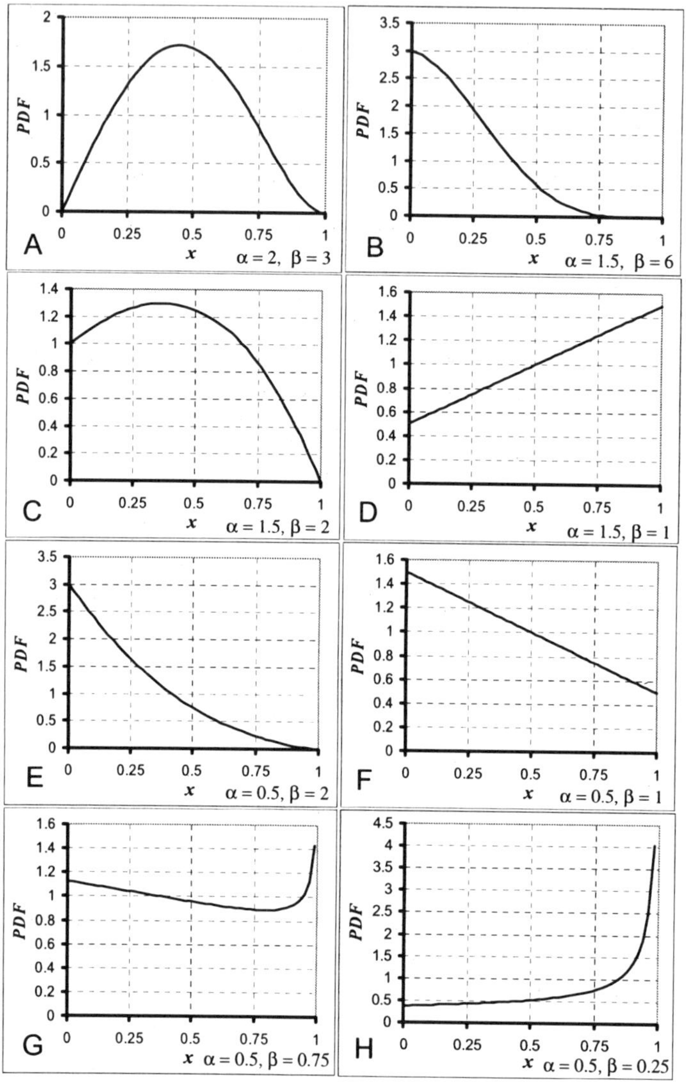

Figure 3. Examples of Standard RGTL distributions ($a = 0$, $b = 1$):
A: $\alpha = 2, \beta = 3$; B: $\alpha = 1.5, \beta = 6$; C: $\alpha = 1.5, \beta = 2$; D: $\alpha = 1.5, \beta = 1$
E: $\alpha = 0.5, \beta = 2$; F: $\alpha = 0.5, \beta = 1$; G: $\alpha = 0.5, \beta = 0.75$; H: $\alpha = 0.5, \beta = 0.25$

Note that in case of Figure 3B the pdf assumes a similar form to that of a reliability function whereas Figure 3C displays a mode at a value greater than 0. Similarly in Figures 3E to 3H the pdf's with the same value of α ($= 0.5$) with progressively decreasing β from 2 to 0.25, indicate the change in form of the pdf from a monotonically decreasing concave form, a linear function with decreasing slope, a mild U-shaped function, up to a monotonically increasing convex curve.

The J-shaped form of the pdf in Figure 3E ($a = 0$, $b = 1$, $\alpha = 0.5$, $\beta = 2$) resembles that of a Weibull distribution with the shape parameter less than one (but on a bounded domain). Note that the structure of (1) is reminiscent to that of the Weibull cdf. Figures 3G and 3H depict a U-shaped pdf form ($a = 0$, $b = 1$, $\alpha = 0.5$, $\beta = 0.75$) and a J-shaped pdf form ($a = 0$, $b = 1$, $\alpha = 0.5$, $\beta = 0.25$) respectively, and are similar to those appearing in the beta family, but with a bounded density value at its lower bound (cf. (3)). Setting $\alpha = 1$, $\beta = 1$ in (2) yields a uniform distribution on $[a, b]$. Hence, analogously to the four parameter beta distribution with the pdf

$$\frac{\Gamma(\alpha+\beta)}{\Gamma(\alpha)\Gamma(\beta)(b-a)}\left(\frac{x-a}{b-a}\right)^{\alpha-1}\left(\frac{b-x}{b-a}\right)^{\beta-1} \tag{6}$$

where $a \le x \le b$, $\alpha > 0$ and $\beta > 0$ and the Two-Sided Power family (see van Dorp and Kotz [3,4]) with the pdf

$$\begin{cases} \dfrac{n}{b-a}\left(\dfrac{x-a}{m-a}\right)^{n-1} & a < x \le m \\ \dfrac{n}{b-a}\left(\dfrac{b-x}{b-m}\right)^{n-1} & m \le x < b \end{cases} \tag{7}$$

where $n > 0$, the RGTL family has the uniform distribution on $[a, b]$ as one of its members. Another common member amongst these 3 families (Beta, TSP and RGTL) is the reflected power (RP) distribution on $[a, b]$ the pdf

$$f(x|\, a, b, \alpha, \beta) = \frac{\beta}{b-a}\left(\frac{b-x}{b-a}\right)^{\beta-1} \tag{8}$$

obtained by substituting $\alpha = 1$ in (2). Substituting $\alpha = 0$ in (2) also yields the reflected power distribution but with parameter 2β. The reader is encouraged to construct diagrams connecting the above cited distributions.

A distinguishing feature amongst RGTL distributions, compared with distributions (6) and (9), is the existence of additional pdf forms with a positive density value at its lower bound (see Figures 3B-3H) allowing representation of uncertain phenomena with such a property. Another feature of RGTL distribution (indicating a lesser flexibility within the same family) is that the pdf's of a GTL distributions and its reflections possess different functional forms, whereas the reflection of a TSP pdf as well as a beta pdf belong to the same functional family.

3. Properties of Standard RGTL distributions

We shall provide some properties of the Standard RGTL (SRGTL) distributions (setting a = 0 and b = 1 in (1) and (2) with the cdf

$$F(x|\,\alpha,\beta)=1-(1-x)^{\beta}\{\alpha-(\alpha-1)(1-x)\}^{\beta} \tag{9}$$

and the pdf

$$\begin{aligned} f(x|\,\alpha,\beta) &= \beta(1-x)^{\beta-1}\times \\ &\{\alpha-(\alpha-1)(1-x)\}^{\beta-1}\{\alpha-2(\alpha-1)(1-x)\} \end{aligned} \tag{10}$$

where $0<\alpha\le 2$ and $\beta>0$. Results may be extended to the general forms of (1) and (2) by means of a simple linear transformation.

Limiting Distributions

It immediately follows from (9) that the pdf (10) converges to a degenerate distribution with a probability mass of 1 at a (b) when $\beta\to\infty$ ($\beta\downarrow 0$) regardless of the value of α.

Stochastic Dominance Properties

Note that for $\beta=1$ (9) simplifies to a slope distribution with the cdf

$$F(x|\,\alpha,\beta=1)=1-\{\alpha(1-x)-(\alpha-1)(1-x)^2\} \tag{11}$$

which is stochastically decreasing in α, i.e.,

$$\alpha_1\le\alpha_2,\; x\in(0,1) \Rightarrow F(x|\,\alpha_1,\beta=1)\ge F(x|\,\alpha_2,\beta=1) \tag{12}$$

Let now $\beta_1\ge\beta_2>0$. From (12) it follows that for all $x\in(0,1)$ and for any β_1

$$1-\{1-F(x|\,\alpha_1,\beta=1)\}^{\beta_1}\ge 1-\{1-F(x|\,\alpha_2,\beta=1)\}^{\beta_1} \tag{13}$$

From the fact that the function z^a is a decreasing function in a for $z\in(0,1)$ it follows from $\beta_1\ge\beta_2>0$ that

$$1-\{1-F(x|\,\alpha_2,\beta=1)\}^{\beta_1}\ge 1-\{1-F(x|\,\alpha_2,\beta=1)\}^{\beta_2} \tag{14}$$

However, simple algebra shows that

$$F(x|\,\alpha,\beta)=1-\{1-F(x|\,\alpha,\beta=1)\}^{\beta} \tag{15}$$

where $F(x|\,\alpha,\beta)$, $F(x|\,\alpha,\beta=1)$ are given by (9) and (11), respectively, which together with (13) and (14) implies

$$\alpha_1\le\alpha_2,\;\beta_1\ge\beta_2,\; x\in(0,1) \Rightarrow F(x|\,\alpha_1,\beta_1)\ge F(x|\,\alpha_2,\beta_2) \tag{16}$$

Hence, RGTL distributions are stochastically increasing in α and stochastically decreasing in β. This seems to be an interesting property shedding an additional light on the meaning of the parameters α and β in (9) and (10), especially in applications. Note that, relation (16) could be verbally expressed as connecting the generating cdf $F(x|\alpha,\beta=1)$ with the generated one, i.e. $F(x|\alpha,\beta)$.

Mode Analysis

As it was already mentioned for $\beta=1$ and $\alpha=1$ the pdf (10) simplifies to a uniform $[0,1]$ density. For $\alpha=1$, $\beta\neq 1$ the pdf (10) becomes a RP distribution (cf. (8)) with a finite mode at 0 with value $\beta>1$ and an infinite mode at 1 for $\beta<1$. Taking the derivative of (10) with respect to x we have

$$\frac{df(x|\alpha,\beta)}{dx}=C(x|\alpha,\beta)f(x|\alpha,\beta) \tag{17}$$

where the multiplier

$$C(x|\alpha,\beta)=$$
$$(\alpha-1)\frac{2}{\alpha-2(\alpha-1)(1-x)}-(\beta-1)\frac{\{\alpha-2(\alpha-1)(1-x)\}}{(1-x)\{\alpha-(\alpha-1)(1-x)\}} \tag{18}$$

is a linear function in β. From the relations

$$\begin{aligned}&f(x|a,b,\alpha,\beta)>0\\&\{\alpha-2(\alpha-1)(1-x)\}>0\\&\{\alpha-(\alpha-1)(1-x)\}>0\end{aligned} \tag{19}$$

for $\alpha\in[0,2]$ and $\beta>1$ it follows from (17) and (18) that the following four additional cases should be considered: Case 1 : $0<\alpha<1$, $\beta\geq 1$; Case 2 : $1<\alpha\leq 2$, $\beta\leq 1$; Case 3 : $1<\alpha\leq 2$, $\beta>1$; Case 4 : $0<\alpha<1$, $\beta<1$

Case 1 : $0<\alpha<1$, $\beta\geq 1$: see figures 3E and 3F :

From (17), (18) and (19) it follows that the SRGTL pdf (10) is strictly decreasing on $[0,1]$ and hence possesses a mode at 0 with the value $\beta(2-\alpha)$ (cf. (3)). For example, setting $\alpha=0.5$ and $\beta=2$ (as in Figure 3E) yields a mode at 0 with value 3. Setting $\alpha=0.5$ and $\beta=1$ (as in Figure 3F) yields a mode at 0 with value 1.5.

Case 2 : $1<\alpha\leq 2$, $\beta\leq 1$: See Figure 3D:

From (17), (18) and (19) it follows that the SRGTL pdf (10) is strictly increasing on $[0,1]$.

From (4) it follows that the pdf (10) has an infinite mode at 1for $\beta<1$ and a finite mode at 1 for $\beta=1$. Setting $\alpha=1.5$ and $\beta=1$ (as in Figure 3D) yields a finite mode at 1 with value 1.5.

Case 3 : $1<\alpha\leq 2$, $\beta>1$: See Figures 3A, 3B and 3C:

This seems to be the most interesting case. From (17), (18) and (19) it follows that the SRGTL pdf (12) may possess a mode in $(0,1)$. Defining $y = 1 - x$ and setting the derivative (17) to zero yields the following quadratic equation in y

$$2\,(\alpha-1)^2 y^2 - 2\alpha(\alpha-1)\,y + \frac{\alpha^2(\beta-1)}{2\beta-1} = 0 \tag{20}$$

(The left hand side of (20) is a parabolic function in y). Noting that the symmetry axis of the parabola associated with the l.h.s. of (20) has the value

$$\frac{\alpha}{2(\alpha-1)} \tag{21}$$

which is strictly greater than 1 for $\alpha > 1$, and that $y = 1 - x \in [0,1] \Leftrightarrow x \in [0,1]$, it follows that out of the two possible solutions of (20) only the solution

$$y^* = \frac{\alpha}{2(\alpha-1)}\left\{1 - \sqrt{\frac{1}{2\beta-1}}\right\} \tag{22}$$

can yield a mode $x^* \in (0,1)$. Moreover, from $1 < \alpha \le 2$, $\beta > 1$ it follows that $y^* > 0$. Also, from (22) we have that $y^* \to \frac{\alpha}{2(\alpha-1)} > 1$ for $1 < \alpha \le 2$ when $\beta \to \infty$. Hence, from (22) we conclude that the mode $x^* = 1 - y^*$ is

$$x^* = \mathrm{Max}\left[0, \frac{1}{2(\alpha-1)}\left\{\alpha\left(1 + \sqrt{\frac{1}{2\beta-1}}\right) - 2\right\}\right] \tag{23}$$

Setting $\alpha = 1.5$ and $\beta = 2$ (as in Figure 3C) yields $x^* = \mathrm{Max}\left[0, -\frac{1}{2} + \frac{1}{2}\sqrt{3}\right] \approx 0.366$. Setting $\alpha = 1.5$ and $\beta = 6$ (as in Figure 3B) yields $x^* = \mathrm{Max}\left[0, -\frac{1}{2} + \frac{3}{22}\sqrt{11}\right] = 0$ and hence a mode is located at the lower bound 0 with value $\beta(\alpha - 2) = 3$ (cf. (3) with a = 0, b= 1). Utilizing (23) it follows that a Standard Reflected Topp and Leone distribution ($\alpha = 2$) has a mode at

$$\sqrt{\frac{1}{2\beta-1}}$$

for $\beta \ge 1$. Setting $\beta = 3$ (as in Figure 3A) yields a mode at $\frac{1}{5}\sqrt{5} \approx 0.447$

Case 4 : $0 < \alpha < 1$, $\beta < 1$: See Figures 3G and 3H:

Similarly to Case 2 it follows that the pdf (10) has an infinite mode at 1for $0 < \alpha < 1$, $\beta < 1$. However, from (17), (18) and (19) it follows that the pdf (10) may also have an anti-mode $x^* \in (0,1)$ (resulting in a U-shaped form) in this case. The formula for the anti-mode is also given by (23) provided $\beta > \frac{1}{2}$. For example, setting $\alpha = 0.5$, $\beta = 0.75$ (as in Figure 3G) yields $x^* = \text{Max}\left[0, \frac{3}{2} - \frac{1}{2}\sqrt{2}\right]$ and hence an anti-mode at approximately 0.793. For $\beta \le \frac{1}{2}$ (as in Figure 3H) the anti-mode of an RGTL distribution occurs at $x^* = 0$, with value $\beta(2-\alpha)$ (cf. (3) with a = 0 , b = 1).

Failure Rate

The failure rate function $r(t) = f(t)/\{1-F(t)\}$ for an SRGTL density follows from (9) and (10) to be

$$D(\alpha, x)\frac{\beta}{1-x} \tag{24}$$

where

$$D(\alpha, x) = \frac{\alpha - 2(\alpha-1)(1-x)}{\alpha-(\alpha-1)(1-x)} \tag{25}$$

and it is straightforward to check that $\beta/(1-x)$ is the failure rate of a standard reflected power (SRP) distribution ((10) with $\alpha = 1$). From (24) it follows that $D(\alpha, x)$ may be interpreted as the relative increase (or decrease) in the failure rate of an SRGTL distribution as compared to a SRP distribution. Taking the derivative of (25) with respect to x yields

$$\frac{\partial D(\alpha, x)}{\partial x} = \frac{\alpha(1-\alpha)}{\{\alpha-(\alpha-1)(1-x)\}^2} \tag{26}$$

Hence, $D(1, x) = 1$ for all $x \in [0,1]$ and it follows from (26) that $D(\alpha, x) < 1$ (>1) for all $x \in [0,1]$ when $1 < \alpha \le 2$ ($0 \le \alpha < 1$). Thus, α may be interpreted as a failure deceleration parameter (relative to the reflected standard power distribution) when $1 < \alpha \le 2$ and a failure acceleration parameter when $0 \le \alpha < 1$. On the other hand, (24) shows that β is a failure acceleration parameter for all $\beta > 0$.

Cumulative Moments

Due to the functional form of the cdf (9) calculations of cumulative moments

$$M_k = \int_0^1 x^k (1-F(x))dx \tag{27}$$

for SRGTL distributions have a slight advantage over that of central moments about the mean. The mean μ_1' and the central moments about the mean μ_2 (variance), μ_3 (skewness) and μ_4 (kurtosis) are connected with the cumulative moments M_k, k = 1,...,4, via

$$\begin{aligned} \mu_1' &= M_0 \\ \mu_2 &= 2M_1 - M_0^2 \\ \mu_3 &= 3M_2 - 6M_1M_0 + 2M_0^3 \\ \mu_4 &= 4M_3 - 12M_2M_0 + 12M_1M_0^2 - 3M_0^4 \end{aligned} \tag{28}$$

(see, e.g., Stuart and Ord [10]). The cumulative moments M_k for SRGTL distributions follow from (9) and (27) to be

$$\begin{aligned} &\int_0^1 x^k (1-x)^\beta \{\alpha - (\alpha-1)(1-x)\}^\beta dx = \\ &= \sum_{i=0}^{k} \binom{k}{i} (-1)^i \alpha^\beta \int_0^1 x^{\beta+i} \left\{1 - \frac{(\alpha-1)x}{\alpha}\right\}^\beta dx \end{aligned} \tag{29}$$

For $\alpha = 1$, expression (29) simplifies to that of the cumulative moments of an SRP distribution (cf. (10) with $\alpha = 1$). For $\alpha \in (1,2]$, the cumulative moments can be expressed utilizing the incomplete Beta function

$$B(x|\alpha,\beta) = \frac{\Gamma(a+b)}{\Gamma(a)\Gamma(b)\int_0^x p^{a-1}(1-p)^{b-1}dp} \tag{30}$$

as

$$M_k = \sum_{i=0}^{k} \binom{k}{i} (-1)^i \alpha^\beta \left\{\frac{\alpha}{\alpha-1}\right\}^{\beta+i+1} \left\{ \frac{B\left(\frac{\alpha-1}{\alpha}\middle|\beta+i+1,\beta+1\right)}{B^{-1}(\beta+i+1,\beta+1)} \right\} \tag{31}$$

where $B(a,b) = \dfrac{\Gamma(a)\Gamma(b)}{\Gamma(a+b)}$ is the Beta function. Numerical routines for evaluating the incomplete Beta function (30) are well known for a long time and are provided in standard PC software such as e.g. Microsoft Excel. However, for $\alpha \in (0,1)$ expression (29) cannot be further simplified and one has to resort to numerical integration. For the moments of the original Topp and Leone [1] distribution (cf. (5)) the cumulative moments were derived by Nadarajah and Kotz [2]. For $\alpha \in (1,2]$, we have for the cumulative moments M_0, M_1, M_2 and M_3

$$
\begin{aligned}
M_0 &= \alpha^{\beta}\left\{\frac{\alpha}{\alpha-1}\right\}^{\beta+1}\left\{\frac{B\left(\frac{\alpha-1}{\alpha}\,\middle|\,\beta+1,\beta+1\right)}{B^{-1}(\beta+1,\beta+1)}\right\}\\
M_1 &= M_0-\alpha^{\beta}\left\{\frac{\alpha}{\alpha-1}\right\}^{\beta+2}\left\{\frac{B\left(\frac{\alpha-1}{\alpha}\,\middle|\,\beta+2,\beta+1\right)}{B^{-1}(\beta+2,\beta+1)}\right\}\\
M_2 &= -M_0+2M_1+\alpha^{\beta}\left\{\frac{\alpha}{\alpha-1}\right\}^{\beta+3}\left\{\frac{B\left(\frac{\alpha-1}{\alpha}\,\middle|\,\beta+3,\beta+1\right)}{B^{-1}(\beta+3,\beta+1)}\right\}\\
M_3 &= M_0-3M_1+3M_2-\alpha^{\beta}\left\{\frac{\alpha}{\alpha-1}\right\}^{\beta+4}\left\{\frac{B\left(\frac{\alpha-1}{\alpha}\,\middle|\,\beta+4,\beta+1\right)}{B^{-1}(\beta+4,\beta+1)}\right\}
\end{aligned}
\tag{32}
$$

Substituting $\alpha = 2$, in (32) yields the mean $\mu_1' = M_0 = 4^{\beta}B(\beta+1,\beta+1)$ of a Standard Reflected Topp and Leone (SRTL) distribution and hence $1-4^{\beta}B(\beta+1,\beta+1)$ is the mean of a Standard Topp and Leone (STL) distribution on $(0,1)$ (see, Nadarajah and Kotz [2]).

Inverse Cumulative Distribution Function

Utilizing the inverse cdf technique random samples from RGTL distributions may straightforwardly be generated. From (9) we derive that $\{1-F^{-1}(z\,|\,\alpha,\beta)\}$, $z \in [0,1]$ is one of the roots of the quadratic equation in y

$$(\alpha-1)\,y^2-\alpha\,y+\sqrt[\beta]{1-z}=0 \tag{33}$$

Noting that (similarly to equation (20)) the symmetry axis associated with the l.h.s. of the quadratic (33) has a value (21) which is strictly larger than 1 for $1<\alpha\le 2$, it follows that out of the two solutions of (33) only the solution

$$\frac{\alpha-\sqrt{\alpha^2-4(\alpha-1)\sqrt[\beta]{1-z}}}{2(\alpha-1)}$$

can yield $\{1-F^{-1}(z\,|\,\alpha,\beta)\}\in[0,1]$. Analogously, it follows that for $0\le\alpha<1$ only the solution

$$\frac{\alpha+\sqrt{\alpha^2-4(\alpha-1)\sqrt[\beta]{1-z}}}{2(\alpha-1)}$$

can result in $\{1-F^{-1}(z\,|\,\alpha,\beta)\}\in[0,1]$. Hence, we have

$$F^{-1}(z\,|\,\alpha,\beta)=\begin{cases}1-\dfrac{\alpha-\sqrt{\alpha^2-4(\alpha-1)\sqrt[\beta]{1-z}}}{2(\alpha-1)} & 1<\alpha\le 2\\ 1-\sqrt[\beta]{1-z} & \alpha=1\\ 1-\dfrac{\alpha+\sqrt{\alpha^2-4(\alpha-1)\sqrt[\beta]{1-z}}}{2(\alpha-1)} & 0\le\alpha<1\end{cases} \tag{34}$$

where the case $\alpha=1$ follows from the cdf of a standard reflected power (SRP) distribution ($\alpha=1$ in (9)).

4. Maximum likelihood estimation

Below we shall discuss an approximate MLE procedure for a total of N observations grouped in m intervals $[x_{i-1},x_i]$ with n_i observations each and interval mean values $\overline{x}_i$, where $x_0\equiv 0$, $x_m\equiv 1$ and

$$N=\sum_{i=1}^{m} n_i$$

The data described above may be summarized in an m-vector x whose elements are the interval mean values and an m-vector n containing the number of observations in each interval. The approximate MLE procedure below may easily be modified to a non-approximate MLE procedure utilizing order statistics, but here our approach is tailored to the format of the income distribution data to be presented in Table 1. The approximate MLE procedure will assume that the probability mass is concentrated at the interval mean $\overline{x}_i$ of the intervals $[x_{i-1},x_i]$. Utilizing (10) we have the likelihood $L(\alpha,\beta\,|\,x,n)$ to be proportional to

$$\beta^N\prod_{i=1}^{m}\left[\left\{\alpha y_i-(\alpha-1)y_i^{\,2}\right\}^{\beta-1}\left\{\alpha-2(\alpha-1)y_i\right\}\right]^{n_i} \tag{35}$$

where

$$y_i=1-\overline{x}_i \tag{36}$$

Instead of maximizing $L(\alpha,\beta\,|\,x,n)$ we may equivalently maximize the log-likelihood. Taking the logarithm of (35) and calculating the derivative with respect to β we obtain

$$\frac{N}{\beta}+\sum_{i=1}^{m} n_i\,\mathrm{Ln}\left\{\alpha y_i-(\alpha-1)y_i^{\,2}\right\} \tag{37}$$

It follows from (37) that

$$\hat{\beta} = N\left[\sum_{i=1}^{m} n_i \, Ln\left\{\frac{1}{\alpha y_i - (\alpha - 1)y_i^2}\right\}\right]^{-1} \tag{38}$$

is the unique MLE of β given a particular value of α. Taking the logarithm of (35) and calculating the derivative with respect to α, one obtains

$$(\beta - 1)\sum_{i=1}^{m} \frac{n_i(1-y_i)}{\alpha - (\alpha - 1)y_i} + \sum_{i=1}^{m} \frac{n_i(1-2y_i)}{\alpha - 2(\alpha - 1)y_i} \tag{39}$$

Substituting (38) into (39) (utilizing $\hat{\beta}$ instead of β and expressing $\hat{\beta}$ in terms of α) the following function $\Psi(\alpha)$ is derived:

$$\Psi(\alpha) = \left[\frac{N}{\sum_{i=1}^{m} n_i \, Ln\left\{\frac{1}{\alpha y_i - (\alpha - 1)y_i^2}\right\}} - 1\right]\sum_{i=1}^{m} \frac{n_i(1-y_i)}{\alpha - (\alpha - 1)y_i} + \sum_{i=1}^{m} \frac{n_i(1-2y_i)}{\alpha - 2(\alpha - 1)y_i} \tag{40}$$

where y_i is given by (36) and the function is defined on a bounded range of $0 \le \alpha \le 2$ .The MLE $\hat{\alpha}$ follows as one of the roots of the equation $\Psi(\alpha) = 0$ or as one of the boundary values $\alpha = 0$ or $\alpha = 2$. The bounded domain of $\Psi(\alpha)$ allows for straightforward plotting of the function in standard spreadsheet software such as Microsoft Excel and subsequent determination of an approximate solution of the MLE $\hat{\alpha}$. Using the root finding algorithm Goalseek, available in Microsoft Excel, and the approximate solution of $\hat{\alpha}$ allows us to calculate $\hat{\alpha}$ up to a desired level of accuracy. Finally, substitution of $\hat{\alpha}$ in to (38) yields the MLE $\hat{\beta}$. The MLE procedure above will be demonstrated in the next section using U.S. 2001 income data.

5. Fitting 2001 U.S. income distribution data

In a leading article of the 459 issue of the Journal of the American Statistical Association (2002, Vol. 97, pp. 663–673) by Barsky et al. [8] an illuminating and comprehensive analysis of the African-American and Caucasian (Non-Hispanic) wealth gap was presented based on a longitudinal survey of approximately over 6000 households over the period 1968 1992. The authors argue that a parametric estimation of the wealth-earning relationship by race is not an appropriate approach. Their main objection is that the wealth-earning relationship is non-linear with an unknown functional form which is difficult to parameterize and parametric estimation may thus likely yield inaccurate estimates. The authors also provide an extensive and up-to-date bibliography up to and including 2001. Barsky et al. [8] note that the racial wealth gap far exceeds the racial income gap at the higher wealth ranges, suggesting that the racial wealth gap is too large to be explained by income gap alone. On the other hand, they conclude that the role of earnings differences is largest at the lower tails of the wealth distribution and decreases dramatically at higher wealth levels. In fact, their results indicate that differences in household earnings account for all of the racial wealth difference in the first quartile of the wealth distribution. Interested readers are also referred to Couch and Daly [9] and O'Neill et al. [11] who study the related topic of the racial wage gap in the U.S.

Our approach to this problem is somewhat different. We attempt to use the distribution developed in the previous sections to fit the more recent household income data in the U.S. for the year 2001 (Source: U.S. Census Bureau, Current Population Survey, March 2002) classified according to the Caucasian (Non-Hispanic), African-American and Hispanic populations and draw some tentative conclusions about the racial income gap based on this data. Parametric estimation of income data is quite common for almost 100 years and a wide variety of distributions have been proposed (see, Kleiber and Kotz [12] for an extensive bibliography). RGTL distributions (which are not discussed in Kleiber and Kotz [12]) allow for a strictly positive density value at its lower bound, which is observed in a non parametric kernel density estimate of the 1989 income data (see Figure 2 of Barsky et al. [8] p. 668). The new distribution we are proposing turns out to be appropriate for the U.S. 2001 household income data, especially for that of the African-American sub population. We emphasize that the main purpose of the numerical analysis below is to illustrate the fitting attributes of the RGTL distribution and properties of its parameters. The numerical analysis herein in no way yields a conclusive answer to the problem of racial income gaps (nor that of racial wealth and racial wage gaps) while providing indications of the current state of affairs and further study is in order.

Table 1 below contains income distribution data for households in the year 2001 for the different ethnic groups: Caucasian (Non-Hispanic), African-American and Hispanic throughout the U.S.A. The MLE procedure above will be used to fit RGTL distributions for incomes of these three groups. Only the data up to \$250,000 will be used in Table 1 since the U.S. Census Bureau data does not provide the maximum observed income in their statistics. Of the total number of U.S. households surveyed 98.58%, 99.44%, 99.65% have in 2001 an income less than \$250,000 for the Caucasian (Non-Hispanic), Hispanic and African-American ethnic groups, respectively.

Figure 4 displays a graph of the function $\varsigma(\alpha)$ (cf. (40)) for the income data of Caucasian (Non-Hispanic) Americans presented in Table 1. From Figure 4 we observe an approximate root of the equation $\varsigma(\alpha)=0$ to be the value $\alpha^* \approx 1.70$. Since, $\varsigma(\alpha)>0$ (<0) for $0 \le \alpha < \alpha^*$ ($\alpha^* \le \alpha \le 2$) it follows that $\hat{\alpha}=\alpha^*$ is the unique MLE of (35) for α. Using Goalseek (a standard root finding algorithm in Microsoft Excel) with a accuracy of 1 . 10^{-6}, utilizing the approximate solution 1.70 we obtain $\hat{\alpha}=\alpha^*=1.679$. The unique MLE $\hat{\beta}=6.767$ follows from substituting $\hat{\alpha}=1.679$ in (38). Figure 5 below plots both the empirical and their fitted RGTL counterparts (cf. (1) and (2)) with a = \$0, b = \$250,000 , $\alpha=1.679$ and $\beta=6.767$. Differences between the empirical cdf and fitted cdf can be observed in Figure 5A. The Kolmogorov-Smirnov Statistic D, which is the maximum observed difference between the empirical and fitted cdf's (see, e.g., DeGroot [13]), in Figure 5A equals 8.60%.

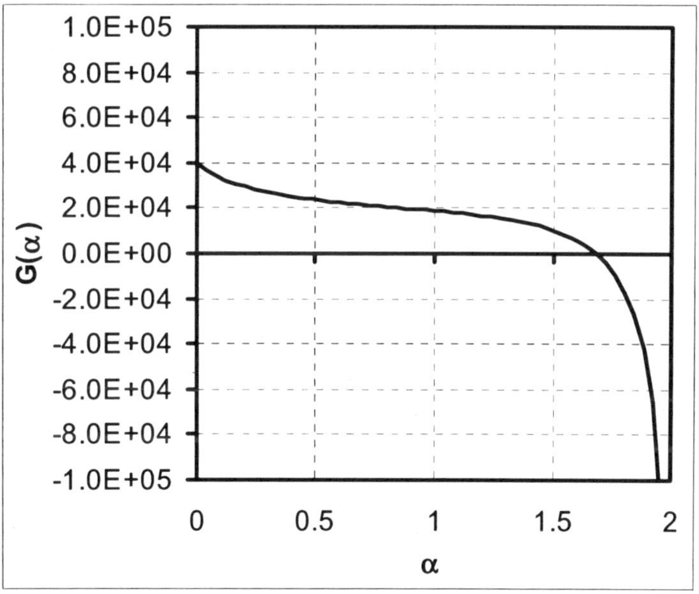

Figure 4. A graph of the function $\varsigma(\alpha)$ (cf. (40)) for the income data of Caucasian (Non-Hispanic) Americans presented in Table 1

Table 1. U.S. income distribution for households in year 2001 (Source: U.S. Census Bureau, Current Population Survey, March 2002. Numbers in thousands, households as of March of the following year)

	Caucasian (Non-Hispanic)		African American		Hispanic	
Income of Household	Number	Mean Income	Number	Mean Income	Number	Mean Income
Under $2,500..............	1,443	$168	520	$439	273	$426
$2,500 to $4,999..........	773	$3,808	388	$3,842	137	$3,767
$5,000 to $7,499..........	2,141	$6,450	698	$6,359	304	$6,387
$7,500 to $9,999..........	2,561	$8,749	756	$8,658	404	$8,663
$10,000 to $12,499.........	3,142	$11,220	621	$11,173	458	$11,214
$12,500 to $14,999.........	2,946	$13,615	543	$13,672	411	$13,659
$15,000 to $17,499.........	3,167	$16,091	660	$16,089	553	$15,993
$17,500 to $19,999.........	2,803	$18,660	479	$18,655	418	$18,579
$20,000 to $22,499.........	3,099	$21,082	610	$21,094	490	$21,005
$22,500 to $24,999.........	2,697	$23,706	447	$23,682	373	$23,691
$25,000 to $27,499.........	3,055	$26,064	570	$26,061	477	$26,011
$27,500 to $29,999.........	2,446	$28,673	464	$28,544	330	$28,617
$30,000 to $32,499.........	3,277	$31,059	492	$31,040	479	$30,998
$32,500 to $34,999.........	2,330	$33,679	375	$33,655	335	$33,601
$35,000 to $37,499.........	2,950	$36,045	437	$35,944	412	$36,082
$37,500 to $39,999.........	2,114	$38,713	310	$38,626	249	$38,641
$40,000 to $42,499.........	2,846	$41,052	434	$41,004	424	$40,938
$42,500 to $44,999.........	1,924	$43,679	260	$43,693	231	$43,668
$45,000 to $47,499.........	2,236	$46,058	289	$45,908	291	$46,044
$47,500 to $49,999.........	1,966	$48,709	256	$48,655	205	$48,607
$50,000 to $52,499.........	2,403	$51,042	350	$50,924	247	$51,021
$52,500 to $54,999.........	1,736	$53,679	210	$53,553	153	$53,725
$55,000 to $57,499.........	2,014	$56,127	249	$55,972	224	$55,992
$57,500 to $59,999.........	1,528	$58,650	177	$58,680	177	$58,764
$60,000 to $62,499.........	2,047	$61,053	248	$60,979	219	$61,106
$62,500 to $64,999.........	1,417	$63,719	162	$63,761	141	$63,801
$65,000 to $67,499.........	1,710	$66,048	175	$65,990	157	$66,018
$67,500 to $69,999.........	1,325	$68,677	150	$68,705	124	$68,734
$70,000 to $72,499.........	1,622	$71,067	190	$71,090	159	$71,112
$72,500 to $74,999.........	1,248	$73,707	142	$73,589	128	$73,711
$75,000 to $77,499.........	1,608	$75,981	133	$75,974	132	$75,860
$77,500 to $79,999.........	1,073	$78,662	100	$78,693	72	$78,726
$80,000 to $82,499.........	1,380	$81,051	100	$80,950	125	$80,976
$82,500 to $84,999.........	993	$83,688	90	$83,584	90	$83,708
$85,000 to $87,499.........	1,144	$86,057	103	$85,984	76	$85,830
$87,500 to $89,999.........	803	$88,696	86	$88,754	55	$88,636
$90,000 to $92,499.........	985	$91,051	78	$91,103	83	$90,997
$92,500 to $94,999.........	701	$93,658	83	$93,666	41	$93,579
$95,000 to $97,499.........	915	$96,071	71	$95,901	65	$95,999
$97,500 to $99,999.........	712	$98,682	65	$98,639	48	$98,811
$100,000 to $149,999.......	8,374	$119,083	554	$117,549	515	$119,016
$150,000 to $199,999.......	2,689	$169,312	115	$172,222	113	$164,692
$200,000 to $249,999.......	993	$219,285	29	$218,672	43	$221,737
$250,000 and above.........	1,345	$462,675	46	$433,097	59	$474,843
Total	**90,682**	**$60,512**	**13,315**	**$39,248**	**10,499**	**$44,383**

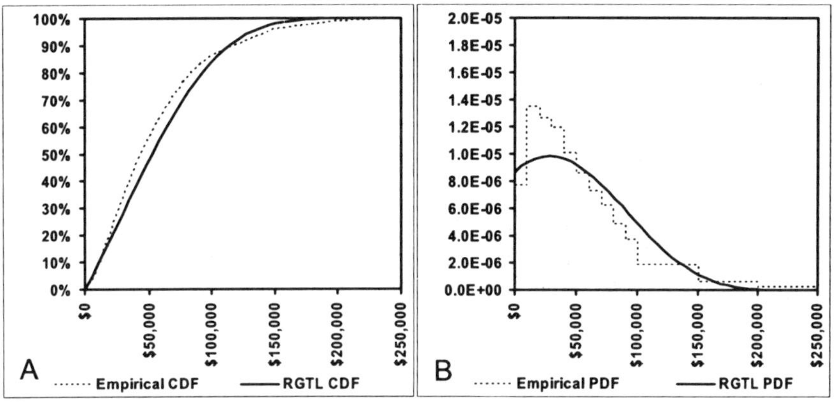

Figure 5. Empirical and an MLE fitted RGTL distribution ($\hat{\alpha} = 1.643$ and $\hat{\beta} = 6.179$) of the Caucasian (Non-Hispanic) income data in Table 1; A: CDF; B: PDF

Hence, with 43 degrees of freedom (Table 1 has 43 rows up to \$250,000) the Kolmogorov-Smirnov test accepts the fitted RGTL distribution at the 10% ($D_{0.10} \approx 0.182$), 5% ($D_{0.05} \approx 0.203$) as well as 1% ($D_{0.01} \approx 0.243$)levels, respectively. Table 2 provides the unique MLE estimators for $\hat{\alpha}$ and $\hat{\beta}$ (obtained using the procedure described in Section 4) for the Caucasian (Non-Hispanic), African-American and Hispanic income data presented in Table 1. Figure 6A (Figure 6B) plots the empirical and fitted RGTL pdf with MLE $\hat{\alpha} = 1.613$, $\hat{\beta} = 10.629$ ($\hat{\alpha} = 1.685$ and $\hat{\beta} = 10.306$) for the African-American (Hispanic) income data as presented in Table 1. The Kolmogorov-Smirnov Statistic D for the African-American (Hispanic) income data equals 6.01% (8.09%) which is smaller than that of the Caucasian (Non-Hispanic) income data (indicating a better fit). Hence the Kolmogorov-Smirnov test accepts both MLE fitted RGTL distributions in Figure 6A and 6B at the 10%, 5% and 1% levels, respectively.

Table 2. Maximum Likelihood Estimators for the parameters $\hat{\alpha}$ and $\hat{\beta}$ of RGTL distributions for the income data in Table 1 up to \$250,000

	$\hat{\alpha}$	$\hat{\beta}$
Caucasian (Non-Hispanic)	1.679	6.767
Hispanic	1.685	10.306
African-American	1.613	10.629

Table 3 contains the (standardized) cumulative moments $M_0 = \mu_1'$, M_1, M_2, M_3 and the central moments μ_2, μ_3 and μ_4 calculated utilizing (32) and (28). Note that there is a strict ordering column-wise for all the values in Table 3 in the order: Caucasian American (non-Hispanic), Hispanic, African-American. From Table 3 we can calculate values for the mean and standard deviation utilizing the transformation Y = \$250,000 X. In a similar manner, the median and mode of the MLE RGTL distributions can be evaluated utilizing the parameter values in Table 2, (34) and (23). In addition, we may utilize Table 3 to calculate the coefficient of skewness β_1 and coefficient of kurtosis β_2 given by

$$\beta_1 = \frac{\mu_3^2}{\mu_2^3}, \quad \beta_2 = \frac{\mu_4}{\mu_2^2}$$

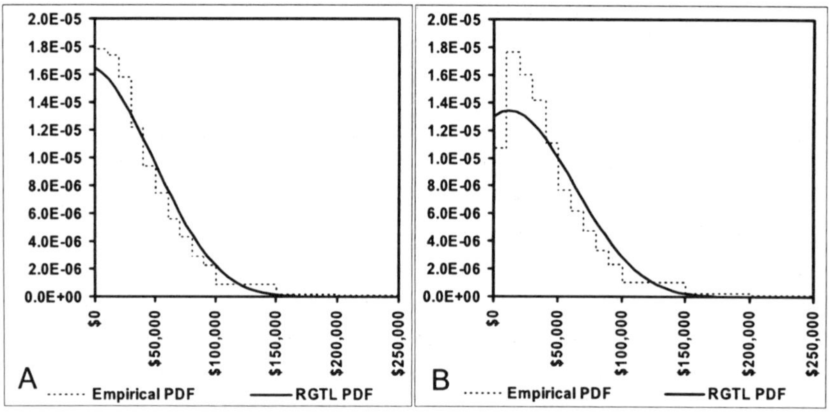

Figure 6. Empirical and MLE fitted RGTL pdf's for the income data in Table 1; A: African-American ($\hat{\alpha} = 1.613$ and $\hat{\beta} = 10.629$); B: Hispanic ($\hat{\alpha} = 1.685$ and $\hat{\beta} = 10.306$)

These estimated statistics are provided in Table 4 for the three subpopulations under consideration.

Table 3. Cumulative Moments M_k and Central Moments μ_{k+1} of the MLE fitted RGTL distributions for the income data in Table 1 up to \$250,000 calculated utilizing (32) and (28), k = 1,...,3

	$M_0 = \mu_1'$	M_1	M_2	M_3	μ_2	μ_3	μ_4
Caucasian (Non-Hispanic)	2.34e-1	3.97e-2	1.11e-2	3.80e-3	2.47e-2	2.54e-3	1.75e-3
Hispanic	1.77e-1	2.38e-2	5.26e-3	1.51e-3	1.60e-2	1.66e-3	8.40e-3
African-American	1.59e-1	1.98e-2	4.17e-3	1.14e-3	1.44e-2	1.60e-3	7.31e-3

Table 4. Statistics associated with the MLE fitted RGTL distributions for the income data in Table 1 up to $250,000

	Mean	Median	Mode	St. Dev	β_1	β_2
Caucasian (Non-Hispanic)	$58393	$52534	$28306	$39326	0.424	2.858
Hispanic	$44316	$38606	$11851	$31710	0.660	3.248
African-American	$39786	$33599	$0	$30002	0.858	3.522

A similar ordering as observed in Table 3 can be observed throughout Table 4. Note that the difference in the point estimates in Table 4 between the Caucasian (Non-Hispanic) population and the African-American Population is approximately $18607 or more and those associated with the Hispanic population $13928 or more. The latter observation is amplified somewhat in Table 4 by the fact that the fitted mean income for the Caucasian (Non-Hispanic) population overestimates the empirical mean (of income up to $250,000)by $3936 whereas the fitted mean income for the African-American (Hispanic) population is overestimated by only $1898 ($2357). Perhaps the most notable difference is the modal income value of $0 for the MLE fitted RGTL distribution for the African-American population while the modal income value for the Caucasian (Non-Hispanic) and Hispanic population have a value substantially larger than zero (and the mode for the Caucasian (Non-Hispanic) population is more than twice that of Hispanics). A similar observation can be made by comparing the RGTL distributions in Figure 5B, 6A and 6B. Finally, from Table 2 and (3) we may evaluate the density values at the lower bound, i.e. $f(0|0,\$250{,}000,\hat{\alpha},\hat{\beta})$, presented in Table 5. Hence, in comparison with Americans of Caucasian origin, African-Americans appear to be approximately 1.9 times as likely and Hispanics 1.5 times as likely, in the year 2001, to have negligible income. It is the fact that our MLE fitted RGTL pdf's may take any positive value at its lower bound, that allows us to reach such a conclusion.

Table 5. Density values at the lower bound of the MLE fitted RGTL distributions for the income data in Table 1 up to $250,000

	$f(0\|0,\$250{,}000,\hat{\alpha},\hat{\beta})$
Caucasian (Non-Hispanic)	8.68e-6
Hispanic	1.30e-5
African-American	1.65e-5

The analysis of our investigations presented below seems to be, in our opinion, of some interest and value. In Figure 7A, we utilize the MLE fitted RGTL income distributions by plotting the percentiles of the African-American and Hispanic income distributions against those of the Caucasian (Non-Hispanic) one using (9), (34) and the corresponding MLE values for $\hat{\alpha}$ and $\hat{\beta}$ in Table 2. For example, from Figure 7A, we observe that approximately 70% (65%) of the African-American (Hispanic) population have less income than the median (50%) of the Caucasian (Non-Hispanic) income distribution. Similar comparisons can be made for other percentiles of the Caucasian (Non-Hispanic) income distribution utilizing Figure 7A. For example, 34%(29%) of the African-American (Hispanic) population earn less than what less than 20% of the Caucasian (Non-Hispanic) population earn (i.e. the 20% percentile of the Caucasian (Non-Hispanic) income distribution). Note that the solid curve in Figure 7A involving the African-American (Hispanic) income distribution is located completely above the unit diagonal which implies stochastic dominance of Caucasian (Non-Hispanic) income over that of the African-American (Hispanic) one. The latter can be directly concluded from the MLE values for $\hat{\alpha}$ and $\hat{\beta}$ in Table 2 and (16) for the African-American and Caucasian (Non-Hispanic) comparison but not the Hispanic and Caucasian (Non-Hispanic) comparison. This shows that the implication arrow in (16) cannot in general be reversed.

In a similar manner, Figure 7B utilizes the MLE fitted RGTL income distributions by plotting the percentiles of the African-American and Caucasian (Non-Hispanic) income distributions against those of the Hispanic one. For example, from Figure 7B, we observe that approximately 56% (37%) of the African-American (Caucasian Non-Hispanic) population have less income than the median (50%) of the Hispanic income distribution. We now conclude from Figure 7B that Hispanic income stochastically dominates the African-American one. The latter conclusion also follows immediately from the corresponding MLE values for $\hat{\alpha}$ and $\hat{\beta}$ in Table 2 and (16). However, we can conclude, once again, only by observation in Figure 7B that Hispanic income is stochastically dominated by Caucasian (Non-Hispanic) income (since the line associated with the Caucasian (Non-Hispanic) now happens to be completely below the unit diagonal). This conclusion, as before, cannot be directly obtained from the corresponding MLE values for $\hat{\alpha}$ and $\hat{\beta}$ in Table 2 and (16).

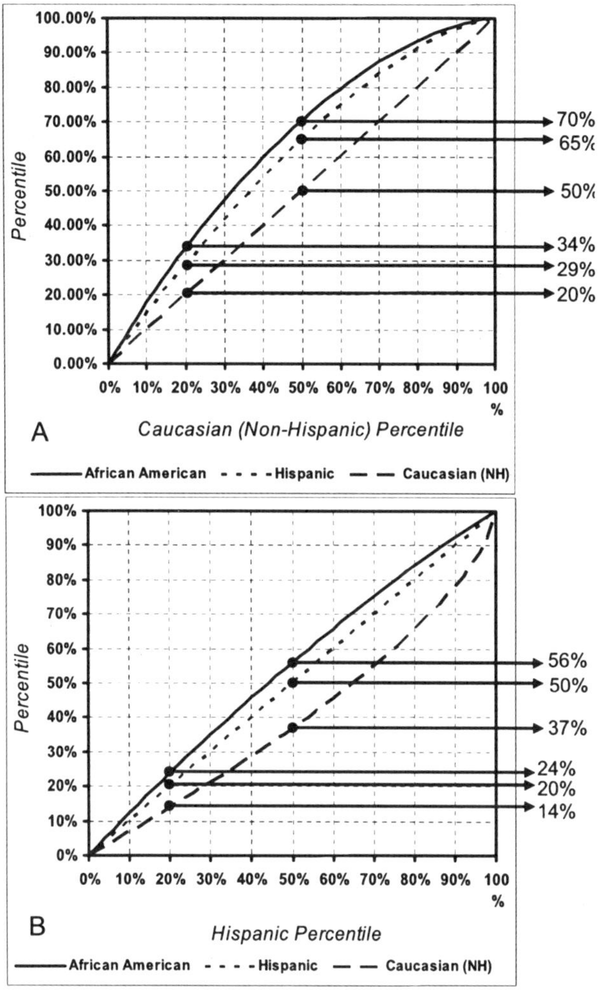

Figure 7. Stochastic Dominance Analysis by Ethnicity for the income data in Table 1 utilizing the MLE fitted RGTL cdf's

Summarizing, Table 2 and (16) alone imply that the chances of a Caucasian (Non-Hispanic) or Hispanic American earning more than a specified amount (anywhere within the range from $0 to $250,000) are higher than those for an African-American. In addition, the analysis in Figure 7 allows us to conclude that the chances of a Caucasian (Non-Hispanic) earning more than a specified amount (anywhere within the range from $0 to $250,000) are higher than those of a Hispanic. Moreover, Figure 7 and Table 4 demonstrate that although substantial advances have reportedly been made in reducing the income distribution gap amongst these three subpopulations in the U.S. during the last 20 years or so (see, e.g., Couch and Daly [9]), these differences still exist and are quite noticeable.

6. Concluding remarks

We have attempted to construct and investigate a new four-parameter continuous family of distributions on a bounded domain possessing arbitrary strictly positive density values at its lower bound. As an illustration, the new family is applied to fitting the distributions of income of Caucasians (Non-Hispanic), Hispanics and African-Americans in the U.S.A. in the year 2001 based on U.S. Census bureau data. The results seems to be quite satisfactory and allow us to compare the incomes of the above 3 groups in a novel manner which seems to be revealing by shedding additional light on features which are not obvious from a direct examination of the raw data.

Acknowledgments

We are indebted to T.A. Mazzuchi for his helpful comments in the course of developing this paper and to Dr. David Findley (U.S. Bureau of Census) for helping us to obtain recent data.

References

1. Topp, C.W. and Leone, F.C. (1955). A family of J-shaped frequency functions. Journal of the American Statistical Association, 50(269), 209–219.
2. Nadarajah, S. and Kotz, S. (2003). Moments of some J-shaped distributions. Journal of Applied Statistics, 30(3), 311–317.
3. Van Dorp, J.R. and Kotz, S. (2002). The standard two sided power distribution and its properties: With applications in financial engineering. The American Statistician, 56(2), 90–99.
4. Van Dorp, J.R. and Kotz, S. (2002). A novel extension of the triangular distribution and its parameter estimation. Journal of Royal Statistical Society, Series D, The Statistician, 51(1), 63–79.
5. Johnson, N.L. (1949). Systems of frequency curves generated by the methods of translation. Biometrika, 36, 149–176.
6. Weibull W. (1939). A statistical distribution of wide applicability. Journal of Applied Mechanics, 18, 293–297.
7. Van Dorp, J.R. and Kotz, S. (2003). Generalized trapezoidal distributions. Metrika, 58(1), 85–97.

8. Barsky, R., Bound, J., Kerwin, K.C. and Lupton, J.P. (2002). Accounting for the Black-White wealth gap: A nonparametric approach. Journal of the American Statistical Association, 97(459), 663–673.
9. Couch, K. and Daly, M.C. (2000). Black-White inequality in the 1990's: A decade of progress. Working Papers in Applied Economic Theory, No. 2000-07, Federal Reserve Bank of San Francisco.
10. Stuart, A. and Ord, J.K. (1994). Kendall's Advanced Theory of Statistics (Vol. 1, Distribution Theory). New York, Wiley.
11. O'Neill, D., Sweetman, O. and Van de Gaer, D. (2002). Estimating counterfactual densities: An application to Black-White wage differentials in the U.S., Economics Department Working Paper Series, Department of Economics, National University of Ireland – Maynooth.
12. Kleiber, C. and Kotz, S. (2003). Statistical Size Distributions in Economics and Actuarial Sciences. New York, Wiley.
13. DeGroot, M.H. (1991). Probability and Statistics, 3rd ed. Reading, MA: Addison-Wesley.

Chapter 2

MAKING COPULAS UNDER UNCERTAINTY

C. GARCIA-GARCIA
Department of Quantitative Methods in Economics
University of Granada
Campus de Cartuja s/n. Granada, 18071, Spain

J.M. HERRERIAS-VELASCO
Department of Quantitative Methods in Economics
University of Granada
Campus de Cartuja s/n. Granada, 18071, Spain

J.E. TRINIDAD-SEGOVIA
Department of Business Administration
University of Almería, Ctra. Sacramento s/n
La Cañada de San Urbano, 04120
Almería, Spain

This paper is based in the MTDF methodology, which lies in obtaining the value of an asset from value of a specific index (Ballestero, 1973). The topics of this paper are to apply this methodology in the case of two indexes under uncertainty, the construction of a copula FGM using marginal TSP given the classical values (a, m, b) and the application in an empirical case. Under an uncertainty environment a high correlation exists between the indexes, which imply the impossibility to apply the FGM copula that is restricted to the weak correlations case. This article overcomes this disadvantage presenting an alternative that later is applied in a practical case.

1. Introduction

The method of the two distribution functions has been developed as a method of valuation recommended under uncertainty environments, this is, when there is no information over the asset that it has to be valued and an experts' is consulted, acting in similar way that in the PERT method.

The present paper is based on the method of the two distribution functions, also known as method of two betas. This method was presented by Ballestero (1971) and it is highly used in valuation. It supposes an improvement of the Synthetic method and it was formalized later by its author, Ballestero (1973), who describes it as follow: the variable market value of a good will follow statistically the distribution function *F*. On the other hand, the index, parameter or explanatory variable will follow statistically another distribution function *G*. We suppose that the functions *F* and *G* have the form of a bell or similar, and then the method of both betas establishes a relationship between both variables.

So, it is necessary to adopt the following hypothesis: if the index L_i of an assets F_i is higher that the L_j of another assets F_j, the market value V_i corresponding to the first assets will be also major that the market value V_j corresponding to the second one.

From it, if the distribution *F* of the market value is known as well as the distribution *G* of the index, the market value V_k corresponding to an index L_k is established by means of the transformation:

$$V_K = \varnothing(L_K) \Leftrightarrow F(V_K) = G(L_K) \tag{1}$$

Palacios, Callejón and Herrerías (2000) have presented a rigorous formalization of the method:

Two random variables related to an asset are considered: the variable *I*, which represents a quality index of the assets, and the variable *V*, asset market value. It is supposed that the value of *V* is a function of its quality, this is to say $V = \varnothing(I)$, where $\varnothing$ is a strictly increasing function in a certain interval $[I_1, I_2]$. If we have a probability distribution whose distribution function is G(i), *V* is a random variable whose distribution function is:

$$F(V) = P[V \le v] = P[\varnothing(I) \le v] = P[I \le \varnothing^{-1}(v)] = G(\varnothing^{-1}(v)) \tag{2}$$

or,

$$G(i) = P[I \le i] = P[\varnothing(I) \le \varnothing(i)] = P[V \le \varnothing(i)] = F(\varnothing(i)),$$

where $\varnothing$ is a strictly increasing function.

It is evident that if F is strictly increasing on the interval $[\varnothing(I_1), \varnothing(I_2)]$, F is invertible on the mentioned interval; from the last expression is obtained $\varnothing(i) = F^{-1}(G(i))$, defined between $(I_1, I_2) \to (\varnothing(I_1), \varnothing(I_2))$ which is a biyection that transforms qualities in market values.

Figure 1 represents two density functions, known the values (*a*, *m*, *b*) and (*a'*, *m'*, *b'*). A biyection remains established between the value of the index and the value of the assets, (García and García, 2003). From here, if the quality index of a good is I_0, then its market value must be:

$$V_0 = \varnothing(I_0) = F^{-1}(G(I_0)) \tag{3}$$

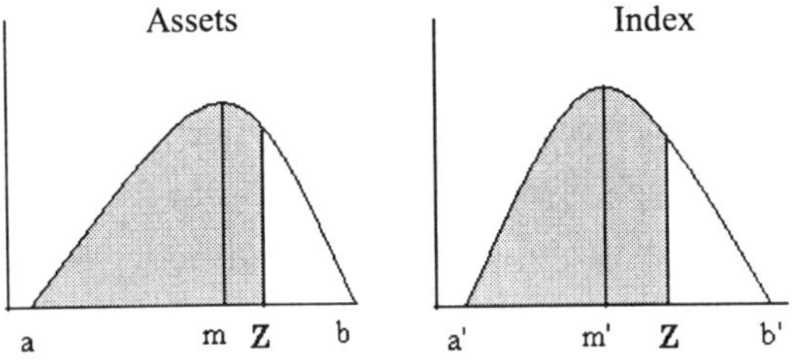

Figure 1. Probability density function in the MTDF for the assets and the index respectively

Since the presentation of the method of the two betas in Ballestero (1973), numerous contributions have been published. These contributions have extended the application of this method and, summarizing, we can distinguish the following lines:

a) Practical applications of the method of the two distribution functions: Ballestero and Caballer (1982), Caballer (1994), Caballer (1998), Caballer (1999) and Ballestero and Rodríguez (1999) extend its use to the valuation of fruit-bearing trees and real estate. Alonso and Lozano (1985) do an application to the valuation of properties in the region of Valladolid; Guadalajara (1996) presents a series of practical cases. García, Trinidad and Sanchez (1997). Cañas, Domingo and Martínez (1994) realize a practical application in the province of Cordoba.

b) Extension of the method to different distributions: Romero (1977) does an extension of the method using uniform and triangular distributions; García, Cruz and Andujar (1998) present a review of the application in triangular distributions. García, Trinidad and Gómez (1999) extend the method to the utilization of a special class of trapezoidal distributions; Herrerias, García, Cruz and Herrerias, (2001) extend the method to the use of trapezoidal distributions of any type. García, Trinidad and García (2004) realize an application using the generalized triangular functions of Van Dorp and Kotz that can be fitted in an uncertainty environment.

c) Utilization of two or more indexes, under the independence hypothesis or not, and implementation of econometric applications. In this line, García, Cruz and Rosado (2000, 2002) present an extension of the method to the multi-index case under the hypothesis of independence between the indexes. Herrerias Velasco (2002) in his Doctoral Thesis extends the method of both distribution functions to the case bivariante of exhaustive form and, in general, to the multivariant case, without hypothesis of independence, and it is also presented the pyramidal distribution. Finally García, Cruz and García (2002.b) present an econometric application of the multi-index case of the two distribution functions method.

d) Development of statistical test to confirm the adequacy of the distribution functions chosen and the kindness of the indexes. García, Cruz and García (2002.a) extend the use of the method of two distribution functions to the mesokurtic functions, constant variance, Caballer and classic betas families introducing a method to select the most appropriated distribution in every case, and presenting a computational program that solves the investment problem. To conclude, Herrerias, Palacios, Callejon and Perez (2001) develop a method to evaluate the kindness of an expert in the PERT methodology.

e) Iteratives procedures of valuation: García, Cruz, García (2002.c) and García, Cruz, García (2004.a).

In this work the MTDF will be applied in the case of two indexes under an uncertainty environment. The FGM family will be used to construct a joint distribution function given the TSP marginal.

2. Initial approach

When one tries to value an asset that depend of one or more indexes and there is no statistical information, it is said that we are in uncertainty environment. The habitual procedure in these cases is to turn to an expert, who will be asked about the optimistic value, the pessimistic value and the most probable value of the assets and, at least, a reference index* (García, Trinidad and García; 2004). If we suppose that we have information about the PERT values for the assets and two reference indexes (see Table 1):

Table 1. PERT values for the assets and two references index

Assets (V)	*Index1 (I_1)*	*Index 2 (I_2)*
(a, m, b)	(a_1, m_1, b_1)	(a_2, m_2, b_2)

* It is been also considered the case of the Multi-Index in García, Cruz and Rosado, (2000, 2002)

The distribution functions $F(I_1)$ and $F(I_2)$ will be obtained from these estimations using classical methods. Then, when the marginal distribution functions are known it is necessary to create a joint distribution function, $F(I_1, I_2)$. This question has been hardly studied in the literature.

The first references are Fréchet (1951) and Lévy (1950), who proved, when he was looking for a definition of the distance between two distributions, that once given a distance between random variables $d(X, Y)$, the minimum of the above mentioned distance, when the distributions of X and Y are given, is another distance. Fréchet, based on Lévy's paper, began to study the problem of creating a joint distribution function when the marginal distributions were known. He proved that from two cumulative distribution function, $F(x)$ and $F(y)$, the joint cumulative distribution function is between W(x,y) and M(x,y):

$$W(x,y) \leq F(x,y) \leq M(x,y) \tag{4}$$

The lower and the upper limits of the previous inequation are usually called Frechet's limits, and these limits are likewise distribution functions with the following expression:

$$\begin{aligned} W(x,y) &= \max\left[F_1(x) + F_2(y) - 1 \ , \ 0\right] \\ M(x,y) &= \min\left[F_1(x) \ , \ F_2(y)\right] \end{aligned} \tag{5}$$

The upper limit is the distribution function of (x,y) when $x = y$ with probability equal to 1, and the lower limit correspond to the case when $y = 1 - x$.

Fréchet was the first in expressing the problem systematically; nevertheless the later develops have been done by other author, mainly Pompilj (1984). It is also interesting the works of Morgenstern (1956), Farlie (1960), Nataf (1962), Plackett (1965), Mardia (1967), Kimeldorf and Sampson (1975). Thesse papers were revised by S. Kotz and N.L. Johnson (1977), but also by Conway (1979) and by Barnett (1980). Schweizer and Sklar (1983) presented the copulas later studied by Genest and Mackay (1986). On the other hand, Johnson and Tenenbein (1989) presented the combined linear weighted method based on Kendall's tau and the Sperman's correlation coefficient. Many of these advances are showed in Dall'Aglio, Kotz and Salinetti (1991). It is observed that this problem has attracted attention in the literature throughout many years, and that still is presented.

There are different topics such us: Fréchet's structure, the joint distribution functions known the marginal distribution functions and the limits, the compatibility of distribution functions, the efficiency of the average value of some functions, the connection with linear programming. It was very important the introduction of copulas by Sklar in 1961, and his later paper with Schweizer, where it was opened a new route of investigation.

3. FGM distribution functions

For many years an interest has existed for the bivariant distribution families known the marginals $F(X_1)$ and $F(X_2)$. These bivariant distributions have usually the following form:

$$F(X_1, X_2 / \alpha) = \psi[F(X_1), F(X_2) / \alpha] \tag{6}$$

where α is a parameter of a vector of parameters.

Some of the most relevant distributions in the literature were introduced by Farlie (1960), Gumbel (1960), Morgensten (1956), Sibuya (1960), Gumbel (1958), Plackett (1965), Frank (1979) and Cook and Johnson (1981). For a discussion on these distributions see Mardia (1970) and Johnson (1987). Genest and Mackay (1986) proved that most of the distributions are obtained from an unique method. Later, Marshall and Olkin (1988) obtains similar conclusions and presents new distributions considering certain mixed models.

In this section we present the FGM distribution function family, which has been widely studied in the literature. The fist reference is Eyraud (1936), who worked in this distribution using uniforms marginals.

The cumulative distribution function of this family has the next form:

$$F(X_1, X_2) = F_1(X_1) F_2(X_2)[1 + \alpha(1 - F_1(X_1))(1 - F_2(X_2))] \tag{7}$$

where:

$F(X_1, X_2)$ is the joint cumulative distribution function of X_1 and X_2.

$F(X_1)$ and $F(X_2)$ are the marginals cumulative distribution functions

The expression for the probability density function is given by:

$$f(X_1, X_2) = f_1(X_1) f_2(X_2)[1 + \alpha(1 - 2F_1(X_1))(1 - 2F_2(X_2))] \tag{8}$$

About the correlation between X e Y it is easy to prove that:

$$E(y / x) = E(y) + \alpha J_2 \{2F_1(x) - 1\} \tag{9}$$

where $J_2 = \int_{-\infty}^{+\infty} F(y)(1 - F(x)) dx$. See Kotz and Drouet (2001).

The α parameter belongs to the interval [−1,1], so the cases in which $\alpha = -1$ and $\alpha = 1$ represent the maximum degrees of negative and positive dependence

respectively, allowed in this family. The dependence properties of this family are associated with the correlation coefficient, though a priori the parameter of the FGM distribution, α, is not associated with this concept.
It is proved that:

- If the marginals follow a *N(0,1)* distribution the correlation is $\alpha\pi^{-1}$, this is equivalent to say that the correlation coeficient moves in the interval *(–0,318, 0,318)*.
- If the marginals follow a uniform distribution, the correlation coeficient is $\alpha/3$ and changes between *–1/3* and *1/3*. It is deduced that for the FGM distribution with absolute continuous marginals, the correlation coeficient between *X* and *Y* can not be higher than *1/3*.

Summarizing, it is possible to affirm that the structural dependence between *X* and *Y* is controlled by the parameter α. To get the density function positive, α has to change between *–1* and *1*. This, restrict the possible values of the correlation coefficient, that changes between *(–1/3,1/3)*, a circunstance that limit the application of the FGM distributions to the cases in which the dependence is weak enough. See Athanassoulis, Skarsoulis, Belibassakis, (1994).

It will be proved later, that under uncertainty environment, due that we have only three data, the existing correlation between the indexes will be out of the range *(–1/3,1/3)* described previously. For it, it is neccesary to try to look for an alternative to apply the family of FGM distribution functions under uncertainty.

4. The Dorp and Kotz's distribution families and its subfamilies

Recently, Van Dorp and Kotz (2002a, 2002b) have introduced the *Two Sided Power* (TSP) distribution, which is a generalization of the triangular distribution and it is defined as follow: Let *x* be a random variable which is said to follow a TSP distribution. Then the probability density function of *x* is:

$$f(x/a,m,b,x)=\begin{cases}\dfrac{n}{b-a}\left(\dfrac{x-a}{m-a}\right)^{n-1}, & \text{si } a<x\le m\\[2ex] \dfrac{n}{b-a}\left(\dfrac{b-x}{b-m}\right)^{n-1}, & \text{si } m\le x<b\end{cases} \tag{10}$$

Standardizing the random variable x, this is, doing the change of variable:

$$t=\frac{x-a}{b-a},$$

We obtain a new random variable t whose density function is given by:

$$f(t/M,n)=\begin{cases} n\left(\dfrac{t}{M}\right)^{n-1}, & \text{si } 0<t\le M \\ n\left(\dfrac{1-t}{1-M}\right)^{n-1}, & \text{si } M\le t<1 \end{cases} \tag{11}$$

The cumulative distribution function is:

$$F(t/M,n)=\begin{cases} M\left(\dfrac{t}{M}\right)^{n}, & \text{si } 0\le t\le M \\ 1-(1-M)\left(\dfrac{1-t}{1-M}\right)^{n}, & \text{si } M\le t\le 1 \end{cases} \tag{12}$$

where:

$$E(t)=\frac{(n-1)M+1}{n+1} \tag{13}$$

and

$$\operatorname{var}(t)=\frac{n-2(n-1)M(1-M)}{(n+2)(n+1)^2} \tag{14}$$

Where a is the pessimistic value, m the most probable value and b the optimistic value, all of them apported by the expert. The parameter n has a more complex interpretation, due that it is not known what exactly means, and also what must be the question asked to the expert to obtain this information. However, we can affirm that n verifies the following properties:

1. $n > 0$
2. For $n = 1$, then the STSP distribution degenerates into a uniform distribution.
3. For $n = 2$, the TSP distribution is tranformed in a triangular distribution with parameters a, m and b.
4. Finally, for $a = 0$ and $m = b = 1$, $f(\cdot/a,m,b,n)$ is a potential function and for $a = m = 0$ and $b = 1$, We would obtain its reflection.

In spite of this, Van Dorp and Kotz point out the intuitive meaning of n since the expected value of x adopts the following expression:

$$E(x) = \frac{a + (n-1)m + b}{n+1}. \tag{15}$$

So, $n - 1$ is the coefficient that wheights the mode to obtain the expected value of the random variable, supposing that the extremes a and b are wheighted by 1. In our opinion this property places the STSP distribution in the field of PERT. From the habitual three classical values a, m and b, whose meaning is known, it should be impossible to determine the unique STSP distribution, since it is a tetraparametric distribution whit parameters a, b, and n. Therefore, it is necessary to restrict the election of the unique STSP distribution to someone of its subfamilies (see García, Cruz and García, 2004.b, 2005).

We will define, first of all, a family of constant variance as the set formed by the STSP distributions with the same variance that the normal distribution of the classic PERT. In case of working with random standardized variables, the following equation is fulfilled:

$$n^3 + 4n^2 + (-72M^2 + 72M - 31)n + (72M^2 - 72M + 2) = 0. \tag{16}$$

This equation allows us to obtain for each value of $M \in (0,1)$, a unique value of $n > 1$. Therefore, we can affirm that, given the three habitual values a, m and b, a unique unimodal STSP distribution with constant variance is set. This result allows to use this family in the field of the PERT. This is, to estimate a TSP, given the values a, m and b.

On the other hand, we will define the mesocurtic family as the set of STSP distributions which kurtosis coeficient (β_2) is equal to 3 (normal kurtosis coefficient). Then the following equation is fulfilled:

$$an^4 + bn^3 + cn^2 + dn + e = 0. \tag{17}$$

where a, b, c, d and e are polinomies in M:

$$\begin{cases} a(M) = 2M^4 - 4M^3 + 6M^2 - 4M + 1 \\ b(M) = -14M^4 + 28M^3 - 22M^2 + 8M - 1 \\ c(M) = -2M^4 + 4M^3 - 22M^2 + 20M - 6 \\ d(M) = 62M^4 - 124M^3 + 94M^2 - 32M + 2 \\ e(M) = -48M^4 + 96M^3 - 56M^2 + 8M \end{cases} \tag{18}$$

It could be proved that for every $M \in (0,1)$, the equation (18) has an only solution that verifies $n > 1$, so we can affirm that always a mesocurtic STSP distribution will exist. This result improves the ones obtained by mesocurtics beta distributions in which is imposible to get a solution when $M \in$ (0,2763933…;0,7236067…). Solving the system created by equations (17) and (18), the unique solutions for $n > 1$ are:

$$\begin{array}{l} M = 0{,}747133\ldots,\ n = 3{,}02344\ldots \\ M = 0{,}252867\ldots,\ n = 3{,}02344\ldots \end{array} \tag{19}$$

These solutions are the same that correspond to the STSP distributions that verify simultaneously the conditions $\sigma^2 = 1/36$ and $\beta_2 = 3$ and these are called classic family.

To conclude, in the PERT, the family of STSP distributions improves always the Beta distributions families due to:

1. It is always possible to select a mesocurtic STSP distribution, while in the case of the Beta distribution it is not possible.
2. As well as the classic Beta distribution exits a STSP distribution with n = 3,0234.
3. The STSP distribution is more moderate in mean and more conservative in variance to every M value. See García, Cruz y García (2005).

This could be explained by the behaviour of the kurtosis coefficient. If we compare the STSP distributions family depending of n and M, with the beta distributions family depending *on* $k = n - 1$ and M, the first one allows to select a distribution with higher kurtosis than the second one. It could be proved that, in symmetric distributions, the value to the kurtosis coefficient of the beta

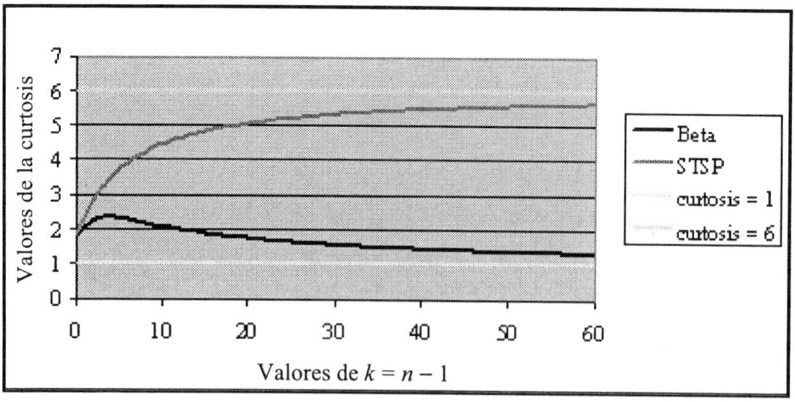

Figure 2. Kurtosis of the Beta and STSP symmetric distributions

distribution is lower that the normal one (3), whereas in the STSP distribution we can find weighted values that get higher or lower kurtosis values than the normal one (see Figure 2). In conclussion, this distribution could be an alternative to the normal distribution and others when we want to fit distributions with a higher kurtosis (see Herrerías, Callejón, Pérez and Herrerías, 2001).

5. An approach to the problem: Application of the MTDF with van Dorp and Kotz's marginals in an uncertainty environment

Once we introduced the FGM distributions families and the TSP distribution, in this section we will obtain the joint distribution funtion applying the FGM distributions family and the STSP distributions. Known two indexes, X_1 and X_2, and the highest, more probable, and lowest values we obtain the following standardized ones:

$$X_1 = \begin{pmatrix} 0 \\ M_1 \\ 1 \end{pmatrix}; \quad X_2 = \begin{pmatrix} 0 \\ M_2 \\ 1 \end{pmatrix} \tag{20}$$

The indexes X_1 y X_2 follow a STSP, so applying the expression (12), its distribution functions are:

$$F(X_1) = \begin{cases} M_1 \left(\dfrac{X_1}{M_1} \right)^{n_1}, & 0 < X_1 < M_1 \\ 1 - (1 - M_1) \left(\dfrac{1 - X_1}{1 - M_1} \right)^{n_1}, & M_1 < X_1 < 1 \end{cases} \tag{21}$$

$$F(X_2) = \begin{cases} M_2 \left(\dfrac{X_2}{M_2} \right)^{n_2}, & 0 < X_2 < M_2 \\ 1 - (1 - M_2) \left(\dfrac{1 - X_2}{1 - M_2} \right)^{n_2}, & M_2 < X_2 < 1 \end{cases} \tag{22}$$

Using (7) the following equation is obtained:

$$F(X_1,X_2)=\begin{cases} M_1\left(\frac{X_1}{M_1}\right)^{n_1} M_2\left(\frac{X_2}{M_2}\right)^{n_2}\left[1+\alpha\left(1-M_1\left(\frac{X_1}{M_1}\right)^{n_1}\right)\left(1-M_2\left(\frac{X_2}{M_2}\right)^{n_2}\right)\right], \\ \qquad 0<X_1<M_1; 0<X_2<M_2 \\ \left[1-(1-M_1)\left(\frac{1-X_1}{1-M_1}\right)^{n_1}\right] M_2\left(\frac{X_2}{M_2}\right)^{n_2}\left[1+\alpha(1-M_1)\left(\frac{1-X_1}{1-M_1}\right)^{n_1}\left(1-M_2\left(\frac{X_2}{M_2}\right)^{n_2}\right)\right], \\ \qquad M_1<X_1<1; 0<X_2<M_2 \\ M_1\left(\frac{X_1}{M_1}\right)^{n_1}\left[1-(1-M_2)\left(\frac{1-X_2}{1-M_2}\right)^{n_2}\right]\left[1+\alpha\left(1-M_1\left(\frac{X_1}{M_1}\right)^{n_1}\right)(1-M_2)\left(\frac{1-X_2}{1-M_2}\right)^{n_2}\right], \\ \qquad 0<X_1<M_1; M_2<X_2<1 \\ \left[1-(1-M_1)\left(\frac{X_1}{M_1}\right)^{n_1}\right]\left[1-(1-M_2)\left(\frac{X_2}{M_2}\right)^{n_2}\right]\left[1+\alpha(1-M_1)\left(\frac{1-X_1}{1-M_1}\right)^{n_1}(1-M_2)\left(\frac{1-X_2}{1-M_2}\right)^{n_2}\right], \\ \qquad M_1<X_1<1; M_2<X_2<1 \end{cases} \tag{23}$$

Applying the expression (8) we get the density funtions:

$$f(X_1,X_2)=\begin{cases} n_1 n_2\left(\frac{X_1}{M_1}\right)^{n_1-1}\left(\frac{X_2}{M_2}\right)^{n_2-1}\left[1+\alpha\left(1-2M_1\left(\frac{X_1}{M_1}\right)^{n_1}\right)\left(1-2M_2\left(\frac{X_2}{M_2}\right)^{n_2}\right)\right], \\ \qquad 0<X_1<M_1; 0<X_2<M_2 \\ n_1 n_2\left(\frac{1-X_1}{1-M_1}\right)^{n_1-1}\left(\frac{X_2}{M_2}\right)^{n_2-1}\left[1+\alpha\left(-1+2(1-M_1)\left(\frac{1-X_1}{1-M_1}\right)^{n_1}\right)\left(1-2M_2\left(\frac{X_2}{M_2}\right)^{n_2}\right)\right], \\ \qquad M_1<X_1<1; 0<X_2<M_2 \\ n_1 n_2\left(\frac{X_1}{M_1}\right)^{n_1-1}\left(\frac{1-X_2}{1-M_2}\right)^{n_2-1}\left[1+\alpha\left(1-2M_1\left(\frac{X_1}{M_1}\right)^{n_1}\right)\left(-1+2(1-M_2)\left(\frac{1-X_2}{1-M_2}\right)^{n_2}\right)\right], \\ \qquad 0<X_1<M_1; M_2<X_2<1 \\ n_1 n_2\left(\frac{1-X_1}{1-M_1}\right)^{n_1-1}\left(\frac{1-X_2}{1-M_2}\right)^{n_2-1}\left[1+\alpha\left(-1+2(1-M_1)\left(\frac{1-X_1}{1-M_1}\right)^{n_1}\right)\left(-1+2(1-M_2)\left(\frac{1-X_2}{1-M_2}\right)^{n_2}\right)\right], \\ \qquad M_1<X_1<1; M_2<X_2<1 \end{cases} \tag{24}$$

Figure 3 represents the joint distribution function given two marginal STSP, by means of the FGM distribution function.

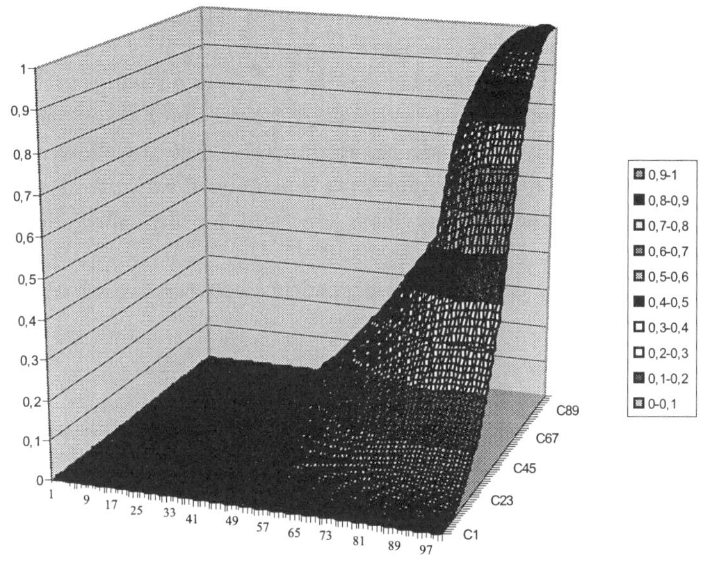

Figure 3. FGM joint distribution function with marginals TSP, $M_1 = 0{,}8$; $M_2 = 0{,}6$; $\alpha = 0{,}9$

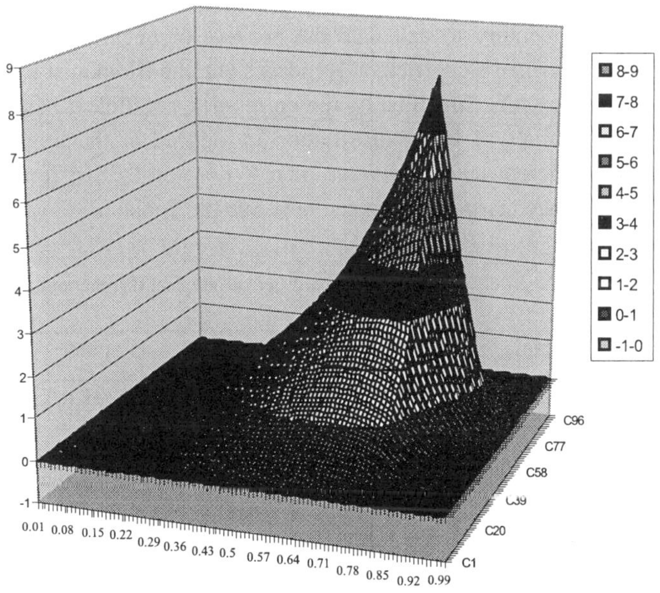

Figure 4. FGM joint density function with marginals TSP, $M_1 = 0{,}8$; $M_2 = 0{,}6$; $\alpha = 0{,}9$

In Figure 4 is presented the FGM joint density function given the Van Dorp marginal functions. To carry out these representations, it is necessary to find before the value of α. Let's remember that the parameter α belongs to the interval [−1,1], so in the cases in which $\alpha = -1$ and $\alpha = 1$ represent the maximum degrees of negative and positive dependence, respectively, that are allowed in FGM family. It is observed that the parameter α is associated with the measurements of dependence, and that is why these are used for the calculation of this parameter.

The basic measurement of linear dependence between two variables X_1 and X_2 is the covariance:

$$\operatorname{cov}(X_1 X_2) = E[(X_1 - E(X_1))(X_2 - E(X_2))] \tag{25}$$

In order that this measurement was independent to the units in which the variables are expressed, the covariance is divided by the product of the standard deviations. This is the widely known correlation coefficient:

$$corr(X_1, X_2) = \frac{\operatorname{cov}(X_1, X_2)}{\sqrt{\operatorname{var}(X_1)\operatorname{var}(X_2)}} \tag{26}$$

This coefficient has been the basic measurement of the linear dependence during more than 100 years. Many other measurements have been proposed during the 20th century to calculate the positive or negative dependence, for example the Spearman's coefficient, Kendall's tau, the Blomquist's q coefficient and the Höffding's Δ. Specifically, the correlation coefficient has been placed for different families of FGM distribution introduced in the literature, so the correlation coefficient takes the value of $\alpha/4$, α/π and $0{,}281\alpha$ for exponential, normal and Laplace marginals, respectively. See the Table 2.

Table 2. Correlation coefficient values given the different marginal distributions in the FGM family

Normal	Uniform	Exponential	Laplace
α/π	$\alpha/3$	$\alpha/4$	$0{,}281\ \alpha$

In these distributions it is not necessary to considerer the value of the parameters of the marginal distributions to calculate the value of the coefficient. Nevertheless, this is not the case of FGM distributions with marginal TSP, since a functional dependence exists on the parameters of the distribution. According to Höffding, W. (1940):

$$\mathrm{cov}(X_1, X_2) = \int_0^1\int_0^1 \{F(X_1, X_2) - F(X_1)F(X_2)\} dX_1 dX_2 \tag{27}$$

Using (7) and (27) we obtain:

$$corr(X_1, X_2) = \alpha \prod_{i=1}^{2} \frac{h_{X_i}}{\sqrt{Var(X_i)}} \tag{28}$$

where:

$$h_{X_i} = \int_0^1 F(X_i)[1 - F(X_i)] dX_i \tag{29}$$

Finally, the substitution of expression (13) into (29) and using (28) and (15), we obtain the following expression for the correlation coefficient:

$$corr(X_1, X_2, M_1, M_2, n_1, n_2) = \alpha \prod_{i=1}^{n} \sqrt{\frac{(n_i + 2)}{n_i - 2(n_i - 1)M_i(1 - M_i)}} \frac{n_i - (n_i - 1)M_i(1 - M_i)}{2n_i + 1} \tag{30}$$

Thus if we observed that, once that the correlation coefficient is known, we might find the parameter α by clearing of the previous expression. To find the correlation coefficient, we will raise the relation between the index *1* and the index *2* as a basic problem of regression. So, one of the indexes will be the explanatory variable (X) and other one the explained one (Y):

$$Y = \beta_0 + \beta_1 X + u.$$

We have three observations (the maximum, most probable and the minimum value) for each of the indexes. The above mentioned values must be standardized. Later we will proceed to realize calculations to obtain the steeming of the parameters and the correlation coefficient:

$$\overline{y} = \begin{pmatrix} 0 \\ M_1 \\ 1 \end{pmatrix}; \quad X = \begin{pmatrix} 1 & 0 \\ 1 & M_2 \\ 1 & 1 \end{pmatrix}.$$

Later, we will operate to find the parameters with the known formula:

$$\hat{\beta} = (X'X)^{-1} X'y$$

where:

$$(X'X)^{-1} = \frac{1}{2(M_2^2 - M_2 + 1)}\begin{pmatrix} M_2^2 & -(M_2'+1) \\ -(M_2'+1) & 3 \end{pmatrix},$$

$$X'\bar{y} = \begin{pmatrix} 1 & 1 & 1 \\ 0 & M_2 & 1 \end{pmatrix}\begin{pmatrix} 0 \\ M_1 \\ 1 \end{pmatrix} = \begin{pmatrix} M_1 + 1 \\ M_2 M_1 + 1 \end{pmatrix}.$$

Solving we obtain:

$$\hat{\beta}_0 = \frac{1}{2(M_2^2 - M_2 + 1)}(M_2^2 + 1)(M_1 + 1) - (M_2 + 1)(M_2 M_1 + 1) = \frac{M_2^2 + M_1 - M_2 - M_2 M_1}{2(M_2^2 - M_2 + 1)}$$

$$\hat{\beta}_1 = \frac{1}{2(M_2^2 - M_2 + 1)}(-(M_2 + 1))(M_1 + 1) + (M_2 M_1 + 1)3 = \frac{2M_2 M_1 - M_2 - M_1 + 2}{2(M_2^2 - M_2 + 1)}$$

The expression for the variances and the covariance can be obtained from Table 3:

Table 3. Calculations

Y	X	$y-\bar{y}$	$x-\bar{x}$	$(y-\bar{y})(x-\bar{x})$
0	0	$-(M_1+1)/3$	$-(M_2+1)/3$	$(M_1+1)(M_2+1)/9$
M_1	M_2	$M_1 - \frac{(M_1+1)}{3}$	$M_2 - \frac{(M_2+1)}{3}$	$(2M_1-1)(2M_2-1)/9$
1	1	$1 - \frac{(M_1+1)}{3}$	$1 - \frac{(M_2+1)}{3}$	$(2-M_1)(2-M_2)/9$

With regard to the variance of *X* and the variance of *Y*:

$$\text{var}(Y) = \left(-\left(\frac{M_1+1}{3}\right)\right)^2 + \left(\frac{2M_1-1}{3}\right)^2 + \left(\frac{2-M_1}{3}\right)^2 = \frac{2}{3}(M_1^2 - 2M_1 + 2) \tag{31}$$

$$\text{var}(X) = \frac{2}{3}(M_2^2 - 2M_2 + 2) \tag{32}$$

Finally, the correlation coefficient is given by:

$$\rho = \frac{(2M_1M_2 - M_1 - M_2 + 2)/3}{\frac{2}{3}\sqrt{(M_1^2 - 2M_1 + 2)(M_2^2 - 2M_2 + 2)}} = \frac{2M_1M_2 - M_1 - M_2 + 2}{2\sqrt{(M_1^2 - 2M_1 + 2)(M_2^2 - 2M_2 + 2)}} \tag{33}$$

If we represent this function in the space we obtain Figure 5:

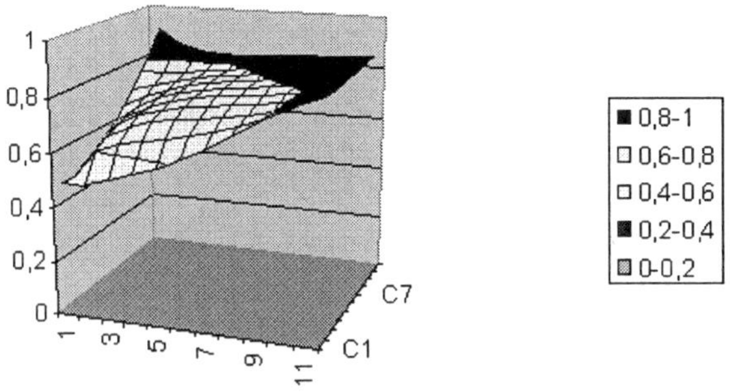

Figure 5. Representation of the correlation coefficient between the index 1 and the index 2 under uncertainty

Once the correlation coefficient is known, it is possible to substitute in the expression (31) and we obtain the value of α. Nevertheless, as it has been observed in the figure (6), the correlation coefficient changes between 0,4986 and 0,9016, and this will give the values of α included in the interval (1,55; 3,16). This fact implies that the expressions (7) and (8) cannot be used since the FGM family only is applicable for values of α between -1 and 1.

In section dedicated to the presentation of this family was warned the disadvantages when the values of α are out of the interval $(-1,1)$, and we conclude that the FGM model is adapted for variables with moderate or small dependence. Nevertheless, in valuation under uncertainty we only have the information contributed by the expert and this make the correlation between the indexes very strong.

6. A solution

Under uncertainty we only have very limited information and this does that the measures of correlation concludes with the existence of a high correlation between the variables. Considering that the parameter α is related to the measures of correlation, it is possible to affirm that the strong existing correlation under uncertainty involves values of α out of the interval $(-1,1)$. This carries the imposibility to apply the family FGM in these cases.

The basic problem is the absence of observations for every index, but if we consider the parameter n as the number of times that the mode is observed, and in this way, n_1 should be the number of times that the mode of the index 1 has be observed and n_2 the number of times that the mode of the second index has been observed, we would possess a total observations of:

$$(n_1+2)(n_2+2) = n_1 n_2 + 2n_1 + 2n_2 + 4$$

Hereby we have passed of having three observations for every index to have $n_1 + 2$ for the first index and $n_2 + 2$ for the second index (see Figure 6). The intention of this, is to avoid high values of the correlation coefficient as a consequence of the absence of information. Nevertheless, the result is that the correlation coefficient will have nule value for every value of n_1 and n_2. It is proposed to omit some of the observations, and it seems to be logical to eliminate that one in which the index 1 takes the optimistic value whereas the index 2 gives us the pessimistic value and vice versa, since these are extreme cases that under a supposition of correlation between the indexes would not be possible. Hereby, the number of observations should be:

$$n_1 n_2 + 2n_1 + 2n_2 + 2 \tag{34}$$

Then,

$$\operatorname{var}(I_1) = \frac{M_1^2\left[2n_1n_2^2 + 6n_1n_2 + 4n_1\right] - M_1\left[2(n_2+1)(n_1n_2 + 2n_1)\right] + \left[n_2^2 + n_1n_2^2 + 3n_1n_2 + 2n_1 + 2n_2 + 1\right]}{(n_1n_2 + 2n_1 + 2n_2 + 2)} \tag{35}$$

$$\operatorname{var}(I_2) = \frac{M_2^2\left[2n_2n_1^2 + 6n_1n_2 + 4n_2\right] - M_2\left[2(n_1+1)(n_1n_2 + 2n_2)\right] + \left[n_1^2 + n_2n_1^2 + 3n_1n_2 + 2n_2 + 2n_1 + 1\right]}{(n_1n_2 + 2n_1 + 2n_2 + 2)} \tag{36}$$

$$\operatorname{cov}(I_1, I_2) = \frac{(M_1 + M_2 - 2M_1M_2)n_1n_2 + (n_1 + n_2 + 1)}{n_1n_2 + 2n_1 2n_2 + 2} \tag{37}$$

We have increased the observations with regard to the initial problem, and we got a nule value of the correlation coefficient. Hereby when the correlation coefficient takes a value into the range $(-1/3, 1/3)$ we will be able to apply the FGM distribution family.

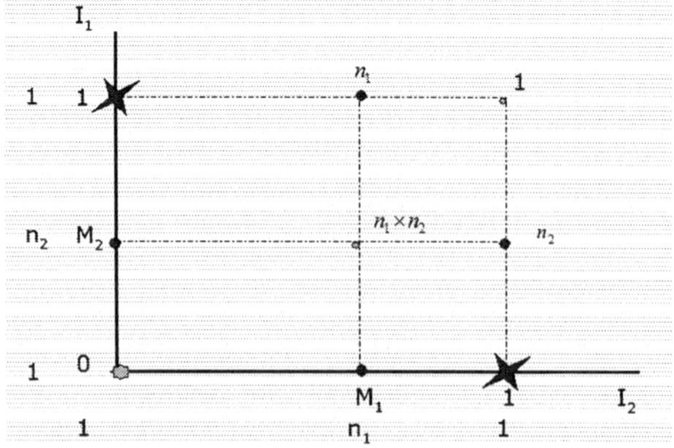

Figure 6. Graph of the proposed solution in which the parameter n is consider as the number of times that the mode of the index is observed

In Figure 7 and Figure 8 it is observed that applying the proposed solution the correlation coefficient should be always minor than 0,3, so the necessary condition to use the FGM distribution family in case of uncertainty would be fulfilled.

We know that:

$$\rho = \frac{Cov(I_1, I_2)}{\sqrt{Var(I_1) \cdot Var(I_2)}}$$

And using the previous expression we obtain Figure 7 in which the correlation coefficient is represented using the expressions (35), (36) and (37).

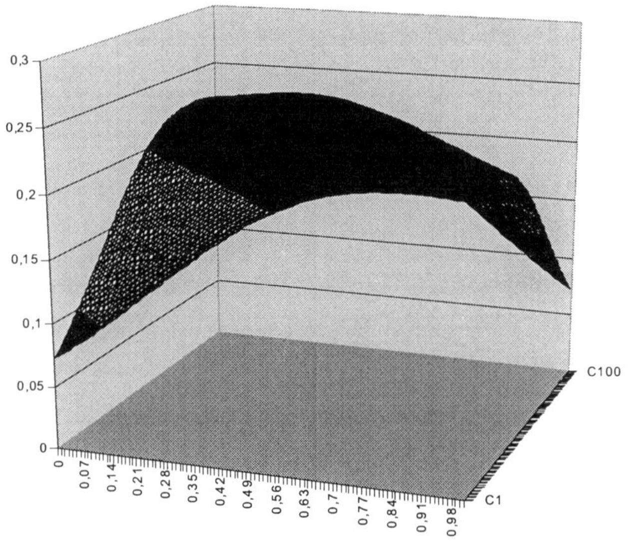

Figure 7. Representation of the correlation coefficient using the proposed solution

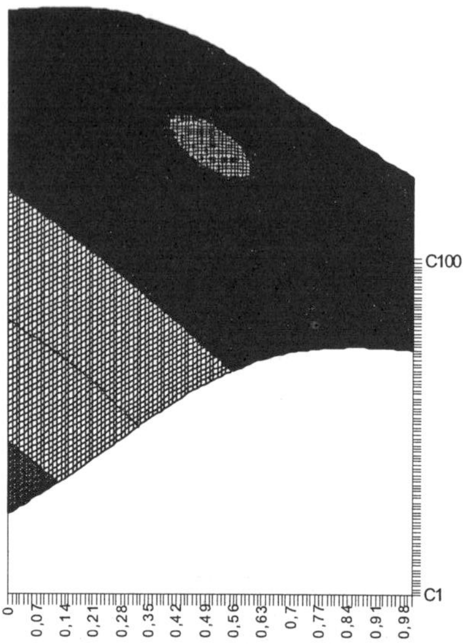

Figure 8. Detail of the representation of the correlation coefficient using the proposed solution

7. A valuation method

Until now it is been developed the procedure to calculate a joint distribution function when the marginal TSP is known, in the case of uncertainty as well as under risk, an expression for it has been achieved. In addition there has been realized a mathematical formulation of the above mentioned procedure.

The next step is to present the valuation procedure that lies in, given values, (x_0, y_0) of both indexs 1 and 2, respectively, to calculate the value of the asset for those values.

First, the value F_0 is calculated for the values (x_0, y_0), this is equivalent to say:

Later, two possibilities are presented depending of the value of F_0, if it is or not major than the standarized mode of the assets *(M)*, and the final value of the assets will depend of this. These posibilities are defined in the expressions (38) and (39):

Branch 1: If $F_0 \prec M$ then:

$$M\left(\frac{V}{M}\right)^n = F_0 \Rightarrow V = M\sqrt[n]{\frac{F_0}{M}} \tag{38}$$

Branch 2: If $F_0 \succ M$ then:

$$1-(1-M)\left(\frac{1-V}{1-M}\right)^n = F_0 \Rightarrow V = 1-(1-M)\sqrt[n]{\frac{1-F_0}{1-M}} \qquad (39)$$

When the application is developed under uncertainty, it will be necessary to realize the procedure of valuation for each of the different subfamilies of the TSP distribution, since the parameter n has a complex interpretation and it is not a known information. Nevertheless, under risk it is possible to estimate the parameter n defining clearly the TSP distribution of each one of the indexes.

Now, two practical applications will be presented, first, under uncertainty and, later, under risk.

8. Practical application of the MTDF with van Dorp and Kotz's marginals under uncertainty enviroment

Table 4 countains the optimistic, more probable and pessimistic values apported by the expert for the assets and the two indexes. In addition we have information about reference values of the indexes. The standardized values are presented in brackets.

Table 4. Inputs for the practical application

	a	*m*	*b*
Assets	110	140	180
	(0)	(0,428)	(1)
Index 1	200	230	300
	(0)	(0,3)	(1)
Index 2	90	120	170
	(0)	(0,375)	(1)

As it was commented in previous section, the parameter n has a difficult interpretation and there has not been defined the question that should be realized to the expert to obtain its value, even if the question exists. For it, in this practical application the different values of n are calculated for each of the TSP distributions subfamilies, using the equations (16), (17), (18), (19) and (20). The above mentioned values are countained in Table 5:

Table 5. Values of *n*

n	Mesocurtic	c-v	classic
assets	3,31885	2,7926	3,02344
index1	3,09726	2,93465	3,02344
index2	3,23222	2,83454	3,02344

Once the values for the parameter *n* and *M* are known, it is possible to substitute in the expression (23) and to obtain the joint distribution function. Analogous substituting in the expression (24) the joint density function can be obtained.

Table 6 contains the values of the correlation coefficient as well as the parameter alpha for each of the subfamilies referred in the practical case:

Table 6. Outputs of the practical application

	Mesocurtic	Constant var
Correlation Coefficient	0,253233397	0,2624621
ALFA	0,43737487	0,453333

It could be observed that, in this case, the correlation coefficient is always lower than 1/3 and therefore we will be able to apply the FGM distribution family to solve the joint distribution function with TSP marginal, since, when the correlation coefficient is lower than the above mentioned value, we will obtain values of alpha in the interval (−1, 1) such as this distributions family needs.

Finally, expressions (38) and (39), that define the two possible branches to calculate the value of the assets once the two values of both indexes are known, are applied. Later, figures (9) and (10) present the functions of the assets value for each of the TSP distributions subfamilies.

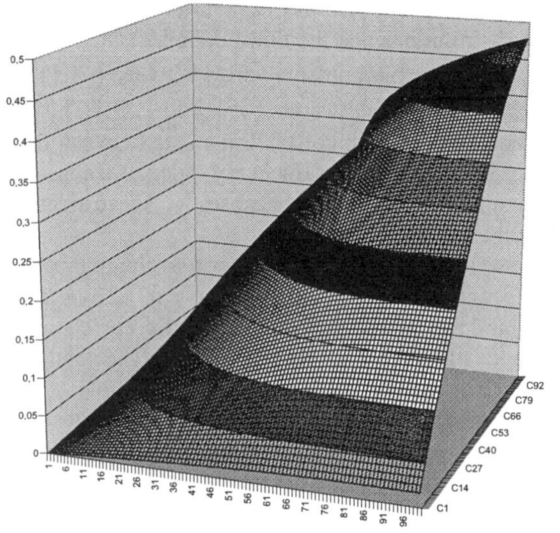

Figure 9. Function of the asset value for constant variante subfamily

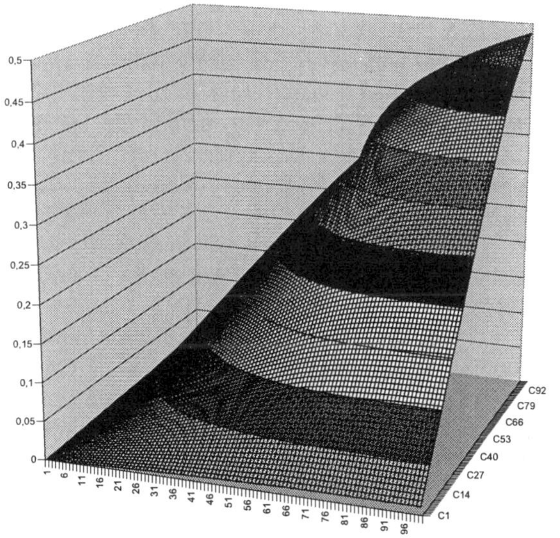

Figure 10. Function of the asset value for mesocurtic subfamily

9. Conclusions

1. Introducing two indexes in the two distributions functions methods under uncertainty we will need the construction of a copula and define

the marginal distributions. We will use the STSP distribution as the marginal distributions and the FGM copula.

2. In FGM copula the parameter α varies between $(-1,1)$ which imply that the correlation coefficient varies between $(-1/3,1/3)$.
3. Under uncertainty a high correlation exists between the indexes due to the absense of observations for every index. As consequence the FGM copula could not be used becauses it is restrected to the weak correlation case.
4. We propose to consider the parameter n as the number of times that the mode is observed. In this way, we have n_i +2 observations for every index I_i, and we avoid high values of the correlation coefficiet as a consequence of the absence of information, The solution is a nule value for the correlation coefficient for every value of n_i.
5. It is proposed to omit extreme cases that under a supposition of correlation between indexes wouln not be possible. It is showed that in this way the correlation coefficient should be always minor than 0,3 so the necessary condition to use FGM distribution family would be fulfilled. In the practical application is showed the valuation method.

References

1. Athanassoulis, G.A., Skarsoulis, E.K. and Belibassakis, K.A. (1994). Bivariate distributions with given marginals with an application to wave climate description. Applied Ocean Research, 16, 1–17.
2. Alonso, R. and Lozano, J. (1985). El método de las dos funciones de distribución: Una aplicación a la valoración de fincas agrícolas en las comarcas Centro y Tierra de Campos (Valladolid). Anales del INIA, Economía, 9, 295–325.
3. Ballestero, E. (1971). Sobre la valoración sintética de tierras y un nuevo método aplicable a la concentración parcelaria. Revista de Economía Política, 225–238.
4. Ballestero, E. (1973). Nota sobre un nuevo método rápido de valoración. Revista de Estudios Agrosociales, 85, 75–78.
5. Ballestero, E. and Caballer, V. (1982). Il metodo delle due beta. Un procedimiento rapido nella stima dei beni fondiari. Genio Rurale, 6, 33–36.
6. Ballestero, E. and Rodriguez, J.A. (1999). El precio de los inmuebles urbanos. CIE Inversiones Editoriales DOSSAT 2000.
7. Barnnet, V. (1980). Some bivariate Uniform Distributions. Communicationes in Statistics, A9, 453–461.
8. Caballer, V. (1994). Métodos de valoración de empresas. Ediciones Pirámide, S.A. 101–104.
9. Caballer, V. (198). Valoración agraria. Teoría y práctica. Ediciones Mundi-Prensa. 4ª edición.

10. Caballer, V. (1999). Valoración de árboles, frutales, forestales, medioambientales, ornamentales. Ediciones Mundi-Prensa.
11. Cañas, J.A. Domingo, J. and Martinez, J.A. (1994). Valoración de tierras en las campiñas y la Subética de la provincia de Córdoba por el método de las funciones de distribución. Investigación Agraria. Serie Economía, 9, 447–467.
12. Conway, D.A. (1979). Multivaríate Distributions with specified Marginals. Technical Report no. 145, Stanford University, Dept. of Statistics.
13. Cook, R. and Jonhnson, M.E. (1981). A family of distirbutions for modelling non-elliptically symetric multivariate data. Journal of the Royal Statistical Society, Series B, 43, 210–218.
14. Dall'Aglio, G., Kotz, S. and Salinetti, G. (1991). Advances in Probability Distributions with Given Marginals. Kluwer Academic Publishers.
15. Eyraud H. (1936). Les principes de la mesure des corrélations. Ann. Univ. Lyon, Sect, A1, 30–47.
16. Farlie, D.J.G. (1960). The perfomance of some correlation coefficients for a general bivariate distribution. Biometrika, 47, 307–323.
17. Frank, M.J. (1979). On the simultaneous associativity of F(x,y) and x+y–F(x,y). Aequationes MATH, 19, 194–226.
18. Fréchet, M. (1951). Sur les Tableux de Corrélation Dont les Marges Sont Dones. Annals de l'Université de Lyon, Ser. 3, 14, 53–77.
19. García, J., Cruz, S. and Andújar, A.S. (1999). Il metodo delle due funzioni di distribuzione: Il modello triangolare. Una revisione. Genio Rurale, 11, 3–8.
20. García, J., Cruz, S. and García L.B. (2002a). Generalización del Método de las dos funciones de distribución (MTDF) a familias betas determinadas con los tres valores habituales. Análisis, Selección y Control de Proyectos y Valoración. Servicio de publicaciones de la Universidad de Murcia.
21. García, J., Cruz, S. and García, L.B. (2002b). Regresión a través de las funciones de Distribución. Actas de la XVI Reunión Asepelt-España, Madrid.
22. García, J. Cruz, S. and García, L.B. (2002c). Iterative valuation process in the meted of the two beta distributions. Spanish Journal of Agricultural Research, 2(1).
23. García, J. Cruz, S. and García, L.B. (2004a). La STSP como distribución subyacente en el ámbito del PERT. Capitulo de libro: Aspectos teóricos y aplicados en la generación de distribuciones de probabilidad. ISBN: 84-931950-8-1.
24. García, J. Cruz, S. and García, L.B. (2004b). Proceso iterativo de valoración en el metodo de las dos betas. Programación, selección, control y valoración de proyectos. Capitulo 3. 37–65. Editado por Universidad de Granada. ISBN: 84-338-3108-9.

25. García, J., Cruz, S. and García, L.B. (2005). The two-sided power distribution for the treatment of uncertainty. Statistical Methods & Applications, 4(5), 209–222.
26. García, J., Cruz, S. and Rosado, Y. (2000). Las funciones de distribución multivariantes en la teoría general de valoración. Actas de la XIV Reunión Asepelt-España, Oviedo (publicación en CD-Rom).
27. García, J., Cruz, S. and Rosado, Y. (2002). Extensión multi-índice del método beta en valoración agraria. Economía Agraria y Recursos Naturales, 2(2), 3–26.
28. García, J. and García, L.B. (2003). Teoria General de valoracion. Método de las dos funciones de distribución. ISBN 84 95979 09 8.
29. García, J., Trinidad, J.E. and García, L.B. (2004). Valoración por el método de las dos funciones de distribución: Como seleccionar la mejor distribución. XVIII Reunion ASEPELT 84-60947165-.
30. García, J., Trinidad, J.E. and Gómez, J. (1999). El método de las dos funciones de distribución: la versión trapezoidal. Revista Española de Estudios Agrosociales y Pesqueros, 185, 57–80.
31. García, J., Trinidad, J.E. and Sánchez, M. (1997). Selección de una cartera de cultivos: el principio primero la seguridad de Roy. Investigación Agraria. Serie Economía, 12(1,2,3), 425–445.
32. Genest, C. and Mackay J. (1986). The joy of copulas: Bivariate distributions with uniform marginals. The American Statistician, 40(4), 280–283.
33. Guadalajara, N. (1996). Valoración Agraria. Casos Prácticos. Ediciones Mundi-Prensa.
34. Gumbel, E.J. (1960). Bivariate exponential distributions. Journal of the American Statistical Association, 55, 698–707.
35. Gumbel, E.J. (1961). Bivariate logistic distributions. Journal of the American Statistical Association, 55, 335–349.
36. Herrerías, R., García, J., Cruz, S. and Herrerías Velasco, J.M. (2001). Il modello probabilistico trapezoidale nel metodo delle due distribución della teoría genérale de valutazioni. Genco Rurale. Estimo e Territorio. Rivista de Scienze Ambientali ANNO LXIV, 4, 3–9.
37. Herrerías, R., Palacios, F., Callejón, J. and Pérez, E. (2001). Un método para contrastar la bondad de un experto en la metodología PERT. Programación, selección y control de proyectos en ambiente de incertidumbre.
38. Herrerías, R., Callejón, J., Pérez, E. and Herrerías, J.M. (2001). Las familias de distribuciones beta de varianza constante y mesocurticas en el método PERT. Programación, selección y control de proyectos en ambiente de incertidumbre.
39. Herrerías Velasco, J.M. (2002). Avances en la teoría general de valoración en ambiente de incertidumbre. Tesis Doctoral.
40. Hoeffding, W. (1940). Maszstabinvariante Korrelationstheorie. Schriften des Mathematischen Instituts un des Instituts für Angewandte Mathematik der Universität Berlin, 5, 181–233.

41. Johnson, Mark E. and Tenenbein A. (1981). A bivariate distributon family with specified marginals. Journal of the American Statistical Association, 76(373), 198–201.
42. Johnson, M.E. (1987). Multivariate Statistical Simulation. New York: John Wiley.
43. Kimeldorf, G. and Sampson, A.R. (1975). One-parameter families of bivariate distributions with fixed marginals. Communications in Statistics, 4, 293–301.
44. Kotz, S. and Johnson, N.L. (1977). On some generalized Farlie-Gumbel-Morgenstern distributions II: Regresion, correlation and further generalization. Communications in Statisics, 6, 415–427.
45. Kotz, S. and Drouet, M. (2001). Correlation and dependence. Imperial College Press.
46. Lévy, P. (1950). Distance de deux variables aléatoires et distance de deux lois de probabilite, in Generalities sur les probabilités. Eléments aléatoires by M. Fréchet, Gauthier-Villars, Paris.
47. Mardia, V. (1967). Some contributions to contingency-type bivariate distributions. Biometrika, 54, 235–249.
48. Marshall, A.W. and Olkin, I. (1988). Families of multivariate distributions. Journal of American Statistical Association, 83, 803–806.
49. Morgenstern, D. (1956). Einfache Beispiele Zweidimensionaler Verteilungen. Mitteinlings fü Mathematische Statistik, 8, 234–235.
50. Nataf, A. (1962). Determination des distributions de Probabilités don des Marges sont Données C.R. Academy of Sciences, 225, 42–43.
51. Palacios, F., Callejón, J. and Herrerias, J.M. (2000). Fundamentos probabilísticos del Método de Valoración de las dos distribuciones. Actas de la XIV Reunión Asepelt-España, Oviedo (publicación en CD-Rom).
52. Placket, R.L. (1965). A class of bivaríate distributions. Journal of the American Statistical Assosiation, 60, 516–522.
53. Pompilj, G. (1984). Le variabili casuali, Ist. Calcolo Probab. Univ. Roma.
54. Romero, C. (1977). Valoración por el método de las dos distribuciones beta: Una extensión. Revista de Economía Política, 75, 47–62.
55. Sibuya, M. (1960). Bivariate extreme statistics. Annals of the Institute of Statistical Mathematics, 19, 195–210.
56. Van Dorp, J.R. and Kotz, S. (2002a). A novel extension of the triangular distribution and its parameter estimation. The Statistician, 51(1), 63–79.
57. Van Dorp, J.R. and Kotz, S. (2002b). The standard two sided power distribution and its properties: With applications in financial engineering. The American Statistician, 56(2), 90–99.
58. Van Dorp, J.R. and Kotz, S. (2003). Generalizations of two sided power Distributions and their convolution. Communications and Statistics: Theory and Method, 32(9).

Chapter 3

VALUATION METHOD OF THE TWO SURVIVAL FUNCTIONS

M. FRANCO-NICOLÁS

Dpto. Estadística e Investigación Operativa, Universidad de Murcia
Campus de Espinardo, Murcia, 30100, Spain

R. HERRERÍAS-PLEGUEZUELO

Department of Quantitative Methods in Economics, University of Granada
Campus de Cartuja s/n. Granada, 18071, Spain

J. CALLEJÓN-CÉSPEDES

Department of Quantitative Methods in Economics, University of Granada
Campus de Cartuja s/n. Granada, 18071, Spain

J.M. VIVO-MOLINA

Dpto. Métodos Cuantitativos para la Economía, Universidad de Murcia
Campus de Espinardo, Murcia, 30100, Spain

In this paper, we discuss a new application of the survival functions in asset pricing from quality indexes. Thus, we propose the valuation method based on the two survival functions (VMTS) to find, under uncertainty, the market value from a quality index. Within this framework, from a one-dimensional quality index, VMTS is equivalent to the valuation method of the two distribution functions (VMTD), which produces loss with respect to the assessments from each component of a multidimensional quality index; nevertheless, VMTS provides profit with respect to these assessments from each component. Finally, we motivate the use of VMTS, as tools for the valuation of an asset, through a practical application on land pricing.

1. Introduction

In the literature, the survival or reliability measures have been widely used in many areas of economics, in political science, in biology, and in industrial engineering. In particular, many interesting results of reliability theory have been applied in risk analysis, and their properties have interesting qualitative implications in these fields (see, e.g. Bagnoli and Bergstrom (2005) and the references therein).

One of these measures is the well-known survival function, also called decumulative distribution function by Yaari (1987), wherein the dual theory of choice under risk is introduced. This paper proposes a new application of this measure in asset pricing from quality indexes.

The asset pricing, under uncertainty, is often analyzed by econometric modelling and hedonic price indexes (see, e.g. Banerjee *et al.* (2004), Deltas and Zacharias (2004) and Benkard and Bajari (2005) and the references therein) as improvements of the classical synthetic method, but the weakness of these techniques is known in the absence of data. In particular, the valuation method based on the two distribution functions (VMTD) has been used to find the market value of an asset, which was introduced by Ballestero(1971) as valuation method of the two beta distributions, and extended by Caballer (1975) and Romero (1977); as well as the practical utility of this methodology when only small samples are available (e.g. Alonso and Lozano (1985)).

Specifically, the beta distribution has been suggested as a rough model in the absence of data, e.g. Law and Kelton (1982), for problems of assessment of risk and uncertainty, such as the program evaluation and review technique (PERT) and the appraisal of an asset from a quality index; so, Berny (1989) proposed a more complicated distribution than the beta model, although their parameters are more intuitive, and Williams (1992) and Johnson (1997) suggested the use of a triangular distribution as a simpler distribution than the beta model, which only requires three parameters (pessimistic, optimistic and most likely).

In this setting, the VMTD allows to appraise an asset under uncertainty, when the appraiser only makes available the pessimistic, optimistic and most likely values, which may be supplied by expert judgement, besides it is simplified to the classical synthetic method when both market value and quality index follow uniform distributions, and it has been used for several applications, such as the valuation of real estates, irrigation, trees, business, ...

In recent years, some authors have paid more attention to the study and generalization of probability models required in PERT methodology and valuation theory (see, e.g. Williams (1992), Johnson (1997), Johnson and Kotz (1999), Herrerías *et al.* (2001), Herrerías (2002), van Dorp and Kotz (2002a), (2002b) and (2003), García and García (2003) and Herrerías *et al.* (2003)), as well as in analysis and development of the VMTD (see, e.g. García *et al.* (1999), Cruz *et al.* (2002), García *et al.* (2002), Herrerías (2002) and García and García (2003)).

In particular, the VMTD has been extended to assess when it is considered a greater information by more than a quality index of the asset (see, e.g. García *et al.* (2002), Herrerías (2002) and García and García (2003)). Unfortunately, the VMTD produces loss with respect to the assessments from each component of a multidimensional quality index, thus it is often used weights among their components to adjust the asset pricing. Therefore, we consider that the new valuation method based on the two survival functions (VMTS) might help to appraise an asset under uncertainty from a quality index, even more when the dimension is reduced by unobserved components of quality.

The purpose of this paper is to establish the theoretical framework of a new valuation method and exhibit its practical application. For that, we study this new technique based on the two survival functions corresponding to the two probability models, providing an in depth explanation of the principles underlying the analysis of the economic value of an asset by means of the VMTS and its comparison with the VMTD.

In Section 2, the VMTD is briefly introduced and the new VMTS is given in order to value an asset, under uncertainty, from a one-dimensional quality index. Section 3 analyzes the VMTS to find the assessment from a bidimensional or multidimensional quality indexes, wherein the differences between both methods are shown when greater information by more than a quality index of an asset is made available. Likewise, in Section 4, the use of the VMTS, as tools for the valuation of an asset, is motivated through a practical application on land pricing, and finally, we provide some concluding remarks in Section 5.

2. Valuation method of the two survival functions

In this section, we introduce a new viewpoint in valuation theory to appraise an asset, under uncertainty, from a one-dimensional quality index, and we establish its relationship with the VMTD in this case. For that, in economic modelling, it is usual to assume certain logical rules of the market. In particular, when we attempt to obtain the market value of an asset from a quality index, the following basic valuation principle is assumed: *the asset with greater quality index has greater market value*, which may be statement as follows:

Let j and k be two assets, with i_j and i_k their values of the quality index and v_j and v_k their market values, respectively. If $i_j < i_k$ then $v_j < v_k$.

Under this assumption, the valuation method of the two distribution functions is based on the equality between the distribution function F_V of the market value V of the asset and the distribution function F_I of its quality index I. Thus, the market value of an asset with quality index $I = i$ by the VMTD is

$$v_D = \phi_D(i) \tag{1}$$

where $\phi_D = F_V^{-1} \circ F_I$.

In this sense, it is possible to consider other valuation techniques where the basic principle holds. In particular, taking into account the survival function S_V of the market value V of the asset and the survival function S_I of a quality index of this asset, instead of their distribution functions, we can consider an alternative method based on the equality of both survival functions, and consequently, the market value of an asset with quality index $I = i$ by this new valuation method of the two survival function is

$$v_S = \phi_S(i) \tag{2}$$

where $\phi_S = S_V^{-1} \circ S_I$, and these survival functions are defined by $S_V(v) = 1 - F_V(v)$ and $S_I(i) = 1 - F_I(i)$. Besides, this VMTS verifies the basic valuation principle, since both survival functions are decreasing.

In order to compare the assessments obtained by Eqs. (1) and (2) through both methods (VMTD and VMTS) from a one-dimensional quality index, we give the following result.

Theorem 1. *Let $I = i$ be the value of the one-dimensional quality index, with v_D its market value by VMTD and v_S its assessment by VMTS. Then, $v_D = v_S$.*

Proof. From Eqs. (1) and (2), it is immediate, since

$$v_S = S_V^{-1}(S_I(i)) = (1 - F_V)^{-1}(1 - F_I(i)) = F_V^{-1}(F(i)) = v_D.$$

Consequently, in the case of an one-dimensional quality index, the VMTS is only a viewpoint alternative to the VMTD. However, in next section, we will see that these methodologies provide different results in the valuation of an asset when more than one quality index is made available to the appraiser.

3. VMTS from a multidimensional quality index

This section provides a comprehensive development of the VMTS when the quality index of the asset is multidimensional, since the professional oftentimes need to determine the value of an asset through a particular set of quality indexes which affect this asset, i.e., using a bidimensional or multidimensional quality index, whose components are each of the one-dimensional quality indexes. For that, we assume the same basic valuation principle, *the asset with greater quality index has greater market value*, where the ordering between two vectors is determined by the orderings between the corresponding components of both vectors.

In this context, we analyze the bidimensional case, and subsequently, we will extend the results to the multidimensional quality index.

3.1.1. *Bidimensional quality index*

In this case, the basic valuation principle can be established as follows:

Let j and k be two assets, with (i_{1j}, i_{2j}) and (i_{1k}, i_{2k}) their values of the quality index and v_j and v_k their market values, respectively. If $(i_{1j}, i_{2j}) < (i_{1k}, i_{2k})$ then $v_j < v_k$.

Analogous to the one-dimensional case, the VMTD is based on the equality between both distribution functions, F_V of the market value and F_I of the bidimensional quality index, and so, the appraisal of an asset with quality index $I = (i_1, i_2)$ by the VMTD is:

$$v_D = \phi_D(i_1, i_2) \tag{3}$$

where $\phi_D = F_V^{-1} \circ F_I$, which provides a market value of the asset lower than the valuations obtained for each component of its quality index, i.e., a reduction or loss in the appraisal of the asset when it is considered a greater information by more than a quality index of this asset, since

$$v_D \leq \inf\{v_1, v_2\} \tag{4}$$

where v_1 and v_2 are the assessments of the asset through each one of the components of the quality index, given by Eq. (1) from their marginal distribution functions F_1 and F_2, respectively.

In this context, the use of an alternative valuation method, wherein the basic valuation principle holds and it equals to the VMTD in the one-dimensional case, is more relevant if it allows us to solve the depreciation in the market of the asset when there are more than a quality index. Specifically, taking into account the survival function S_V of the market value V of the asset and the bivariate survival function S_I of its quality index, the VMTS is based on the equality between both survival functions, and consequently, the assessment of an asset with quality index $I = (i_1, i_2)$ by the VMTS is

$$v_S = \phi_S(i_1, i_2) \tag{5}$$

where $\phi_S = S_V^{-1} \circ S_I$ and the bivariate survival function is defined as:

$$S_I(i_1, i_2) = P(I > (i_1, i_2))$$

which is determined by the bivariate distribution function of the quality index and their marginal distributions

$$S_I(i_1, i_2) = F_I(i_1, i_2) - F_1(i_1) - F_2(i_2) + 1. \tag{6}$$

Likewise, this alternative methodology could say dual of the VMTD, is a new viewpoint to deal the market value of an asset with more than a quality index, which does not lead to loss in the appraisal of the asset. Moreover, the VMTS produces an appraisal of the asset upper than the valuations obtained for each component of its quality index, i.e., an appreciation when more than one quality index is made available to value the asset, as we prove in the next result.

Theorem 2. *Let $I=(i_1,i_2)$ be the value of the bidimensional quality index, with v_1 and v_2 its market values by VMTS from the components I_1 and I_2, respectively. Let v_S be the assessment by VMTS, then, $v_S \geq \sup\{v_1,v_2\}$.*

Proof. Taking into account that for all bivariate survival function the following inequalities holds:

$$S_I(i_1,i_2) \leq S_1(i_1) \text{ and } S_I(i_1,i_2) \leq S_2(i_2)$$

where $S_j(i_j)=1-F_j(i_j)$ for $j=1,2$, are the marginal survival functions of each component of the quality index, i.e.

$$S_I(i_1,i_2) \leq \inf\{S_1(i_1),S_2(i_2)\}$$

and the market values v_1 and v_2 given by Eq. (2) for each component, we have the following inequality

$$v_S \geq \sup\{v_1,v_2\}$$

since S_V is decreasing, and also S_V^{-1}.

However, we establish the comparison between both methods from a bidimensional quality index. So, when we have great information through more quality characteristics of the asset, the appraisals by the VMTS are greater than ones obtained by the VMTD.

Theorem 3. *Let $I=(i_1,i_2)$ be the value of the bidimensional quality index, where v_D is its market value by VMTD and v_S is its assessment by VMTS. Then, $v_D \leq v_S$.*

Proof. Obvious from Theorem 2, the inequality (4) and the equivalence between both methods in the one-dimensional of Theorem 1.

3.1.2. *Multidimensional quality index*

Let us see now, the extension of the VMTS to the multidimensional quality index $I=(I_1,\ldots,I_n)$ with $n \geq 2$, under the assumption of the basic valuation principle, which may be statement as follows:

Let j and k be two assets, with $(i_{1j},...,i_{nj})$ and $(i_{1k},...,i_{nk})$ their values of the quality index and v_j and v_k their market values, respectively. If $(i_{1j},...,i_{nj}) < (i_{1k},...,i_{nk})$ then $v_j < v_k$.

So, from a value of the quality index $I = (i_1,...,i_n)$ of the asset, the VMTD provides the next assessment:

$$v_D = \phi_D(i_1,...,i_n) \tag{7}$$

where $\phi_D = F_V^{-1} \circ F_I$, which is lower than the valuations obtained for each of its components, since

$$v_D \le \inf\{v_1,..., v_n\} \tag{8}$$

where v_j's are the assessments of the asset through marginal distributions F_j's by Eq. (1), $j = 1,...,n$.

Likewise, using the survival function S_V of the market value V of the asset and the multivariate survival function S_I of its quality index, the assessment obtained by the VMTS from $I = (i_1,...,i_n)$, is

$$v_S = \phi_S(i_1,...,i_n) \tag{9}$$

where $\phi_S = S_V^{-1} \circ S_I$.

Analogous to the bidimensional case, the VMTS solves the depreciation underwent by the VMTD when more than one quality index is made available to value the asset, and provides appraisals of the asset greater than the best valuation obtained for each component of its quality index.

Theorem 4. *Let $I = (i_1,...,i_n)$ be the value of the multidimensional quality index, with v_j's its market values by VMTS from the components I_j's, respectively. Let v_S be the assessment by VMTS, then, $v_S \ge \sup\{v_1,..., v_n\}$.*

Theorem 5. *Let $I = (i_1,...,i_n)$ be the value of the multidimensional quality index, where v_D is its market value by VMTD and v_S is its assessment by VMTS. Then, $v_D \le v_S$.*

4. Practical application

In this section, we apply the new valuation method based on the two survival functions to appraise agricultural plots, obtaining the assessments using both valuation methods, wherein one can check the results established in the former sections.

For that, we use the 2nd practical case of Guadalajara (1996), in which she studies the valuation of an agricultural plot, used for growing grapes, in *Vinalopó Medio* region (Alicante, Spain). The quality indexes considered to describe the market value (€/ m^2) are the gross production of grapes (kg/m^2), together with the percentage of sand in the soil of the plot.

Table 1 displays data of the minimum (pessimistic), maximum (optimistic) and mode (most likely) values for each variable; where the goal is to valuate a plot of agricultural land with an area of 12010.3833 m^2 , a gross production of 20399 kg/m^2 and a sand/soil content of 32%.

Table 1. Agricultural plots used for growing grapes

	Pessimistic	Optimistic	Most likely
V=Market value (€/m^2)	0.8132	1.5012	1.0634
I_1=Gross production (kg/m^2)	1.5611	2.6019	1.8734
I_2=Sand/soil content (%)	15	50	25

In order to apply and compare the assessments of this agricultural plot, using the VMTS with the ones obtained by the VMTD from the bivariate probability model of the quality index, we will consider that this quality index follows a pyramidal model and the market value has either triangular or trapezoidal model, which are a sample of the different models that might be considered both in the market value as well as in the bidimensional quality index. Note that to choose the four basic trapezoidal parameters, from these three parameters (pessimistic, optimistic and most likely), we use the specification of the modal interval given in Callejón *et al.* (1996)

Table 2 displays the assessments in both models of the market value, which are multiplied by the 12010.3833 m^2 of the particular plot; in each and every cases, the VMTS allows us to get assessments which reduce loss underwent by the VMTD with respect to the ones obtained for each component of the quality index, as well as the remaining comparisons of Section 3.

Table 2. Assessments from pyramidal quality index

F_V	V_D	$\inf\{V_1, V_2\}$	$\sup\{V_1, V_2\}$	V_S
Triangular	12587.1	13539.2	13747.4	15033.8
Trapezoidal	12773.2	13785.0	13983.3	15198.9

Likewise, it is understandable to think that the gross production of grapes is related to the sand content of the soil, although this correlation may not be very strong. In any way, if we assume independence between the two components of the quality index and both have a triangular model, and the market value follows either triangular or trapezoidal models, we give in Table 3 the assessments of this particular property; where one can also check the comparisons established in Section 3.

Table 3. Assessments from independent and triangular components

F_V	V_D	$\inf\{V_1, V_2\}$	$\sup\{V_1, V_2\}$	V_S
Triangular	12787.3	13775.7	14018.9	15441.2
Trapezoidal	12994.2	14010.0	14239.9	15583.8

Nevertheless, to give a more clear exposition of the behaviour among these assessments, using both VMTS and VMTD from the bidimensional quality index, and a better comparison with respect to the appraisals obtained for each one of the components, we show some graphs corresponding to triangular model for the market value of this agricultural plot, in both bivariate probability models of its quality index (pyramidal or independent and triangular components). Note that to make easy the interpretation of these graphs in the plane, both components of the quality index have been simultaneously taken from the least to the supreme into their supports.

Figure 1 displays the assessments when the quality index has a pyramidal model and Figure 2 depicts the appraisals when the quality index has independent and triangular components, wherein one can see the inequalities of Theorems 2 and 3.

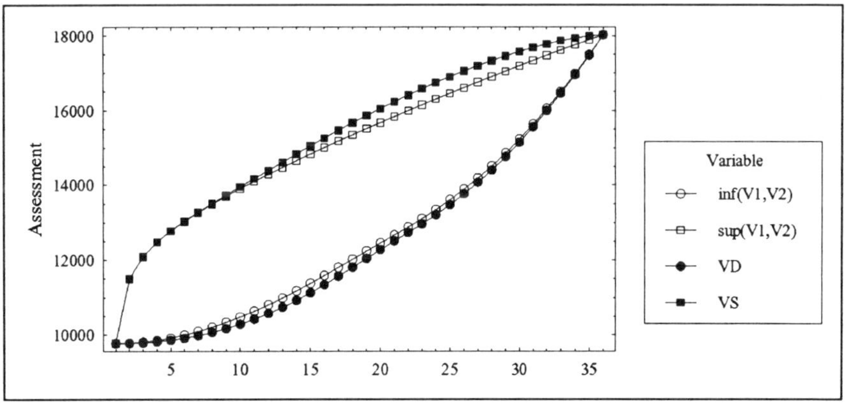

Figure 1. Assessments from pyramidal quality index

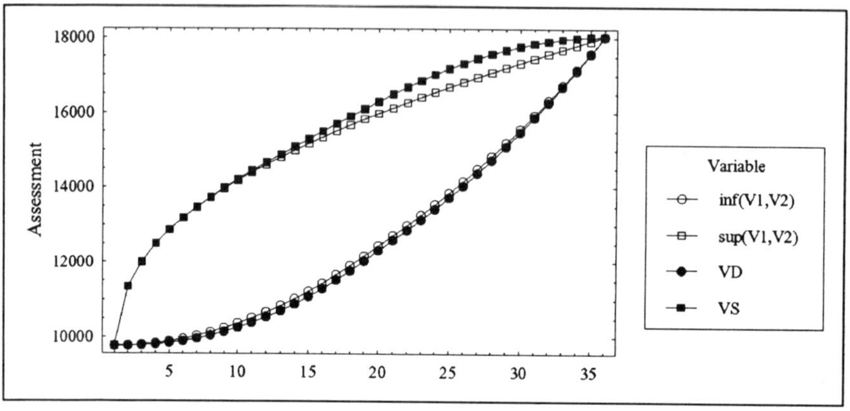

Figure 2. Assessments from independent and triangular components

5. Conclusions

Finally, we point out the main conclusions of this paper.

The VMTD and VMTS are equivalent to appraise an asset from an one-dimensional quality index.

The VMTS provides an greater assessment than one obtained by the VMTD, from a bidimensional or multidimensional quality index. Moreover, the VMTS produces profit with respect to both assessments given by each component of the quality index, solving the depreciation of the VMTD when more than a quality index made is available.

References

1. Alonso, R. and Lozano, J. (1985). El método de las dos funciones de distribución: Una aplicación a la valoración de fincas agrícolas en las comarcas Centro y Tierra de Campos (Valladolid). Anales del INIA: Economía, 9, 295–325.
2. Ballestero, E. (1971). Sobre valoración sintética de tierras y un nuevo método aplicable a la concentración parcelaria. Revista de Economía Política, 57, 225–238.
3. Bagnoli, M. and Bergstrom, T. (2005). Log-concave probability and its applications. Economic Theory, 26, 445–469.
4. Banerjee, A., Gelfand, A.E., Knight, J.R. and Sirmans, C.F. (2004). Spatial modeling of house prices using normalized distance-weighted sums. Journal of Business and Economic Statistics, 22, 206–213.

5. Benkard, C.L. and Bajari, P. (2005). Hedonic price indexes with unobserved product characteristics, and application to personal computers. Journal of Business and Economic Statistics, 23, 61–75.
6. Berny, J. (1989). A new distribution function for risk analysis. Journal of the Operational Research Society, 40, 1121–1127.
7. Caballer, V. (1975). Concepto y métodos de valoración agraria. Ed. Mundi-Prensa, Madrid.
8. Callejón, J., Pérez, E. and Ramos, A. (1996). La distribución trapezoidal como modelo probabilístico para la metodología PERT. In Programación, selección y control de proyectos en ambiente de incertidumbre, R. Herrerías (ed) (2001), 167–177.
9. Cruz, S., García, C.B. and García, J. (2002). Statistical test for the method of the two distribution functions. An application in finance. In VI Congreso de Matemática Financiera y Actuarial and 5th Italian-Spanish Conference in Financial Mathematics, Valencia.
10. Deltas, G. and Zacharias, E. (2004). Sampling frequency and the comparison between matched-model and hedonic regression price indexes. Journal of Business and Economic Statistics, 22, 206–213.
11. García, J., Cruz, S. and Andújar, A.S. (1999). Il metodo delle due funzioni di distribuzione: Il modello triangolare. Una revisione. Genio Rurale, 11, 3–8.
12. García, J., Cruz, S. and Rosado, Y. (2002). Extensión multi-índice del método beta en valoración agraria. Economía Agraria y Recursos Naturales, 2, 3–26.
13. García, J. and García, L.B. (2003). Teoría General de Valoración. Método de las dos funciones de distribución. Ed. Fundación Unicaja, Málaga.
14. Guadalajara, N. (1996). Valoración Agraria. Casos Prácticos. Ed. Mundi-Prensa, Madrid.
15. Herrerías, J.M. (2002). Avances en la Teoría General de Valoración en Ambiente de Incertidumbre. PhD Dissertation, Universidad de Granada.
16. Herrerías, R., García, J. and Cruz, S. (2003). A note on the reasonableness of PERT hypotheses. Operations Research Letters, 31, 60–62.
17. Herrerías, R., García, J., Cruz, S. and Herrerías, J.M. (2001). Il modello probabilistico trapezoidale nel metodo delle due distribuzione della teoria generale de valutazioni. Genio Rurale. Rivista di Scicienze Ambientali, LXIV, 3–9.
18. Johnson, D. (1997). The triangular distribution as a proxy for the beta distribution in risk analysis. Journal of the Royal Statistical Society, Ser. D, 46, 387–398.
19. Johnson, N.L. and Kotz, S. (1999). Non-smooth sailing or triangular distributions revisited after some 50 years. Journal of the Royal Statistical Society, Ser. D, 48, 179–187.
20. Law, A.M. and Kelton, W.D. (1982). Simulation modelling and analysis. Ed. New York: McGraw-Hill.

21. Romero, C. (1977). Valoración por el método de las dos distribuciones beta: una extensión. Revista de Economía Política, 75, 47–62.
22. van Dorp, J.R. and Kotz, S. (2002a). The standard two sided power distribution and its properties: with applications in financial engineering. The American Statistician, 56, 90–99.
23. van Dorp, J.R. and Kotz, S. (2002b). A novel extension of the triangular distribution and its parameter estimation. Journal of the Royal Statistical Society, Ser. D, 51, 63–79.
24. van Dorp, J.R. and Kotz, S. (2003). Generalized trapezoidal distributions. Metrika, 58, 85–97.
25. Williams, T.M. (1992). Practical use of distributions in network analysis. Journal of the Operational Research Society, 43, 265–270.
26. Yaari, M. (1987). The dual theory of choice under risk. Econometrica, 55, 95–115.

Chapter 4

WEIGHTING TOOLS AND ALTERNATIVE TECHNIQUES TO GENERATE WEIGHTED PROBABILITY MODELS IN VALUATION THEORY

M. FRANCO-NICOLÁS

Dpto. Estadística e I.O., Universidad de Murcia
Campus de Espinardo, Murcia, 30100, Spain

J.M. VIVO-MOLINA

Dpto. Métodos Cuantitativos para la Economía, Universidad de Murcia
Campus de Espinardo, Murcia, 30100, Spain

In risk analysis, different procedures based on weighted probability models are usual tools to reduce loss of the assessments in multivariate scenarios. In particular, the weighted distribution functions have been widely used to correct and fit the market value of an asset, through the valuation methods of the two functions, with respect to the appraisals from each component of the multidimensional quality index, in the field of the Valuation Theory.

In this context, the weighting procedures are of interest to find the weights and consequently, to generate these weighted probability models.

The main objective of this paper is to analyze the different weighting techniques used in the Valuation Theory, as well as to propose an alternative to calculate the weights and a new tool to generate these weighted probability models.

First, the well-known weighting techniques to generate the weights are introduced, under both independence and dependence presence of the components of the quality index. Secondly, we expand these weighting techniques by the survival functions, which allows us to generate other weighted probability models.

Likewise, we discuss a new tool to determine the weights of the components of the quality index, modal mean technique, based on the mode values of its marginal distribution functions, which extend the size of the possible weighted probability models to approach the market value of the asset.

Finally, we give an application of these weighting techniques to generate weighted probability models in one example of land pricing, and thus we obtain the assessments of the land property according to each weighted probability model.

1. Introduction

In recent years, some authors have paid more attention to the study and generalization of probability models required in PERT methodology and Valuation Theory (see, e.g. Williams (1992), Callejón, Pérez and Ramos (1996),

Johnson (1997), Johnson and Kotz (1999), Herrerías, García, Cruz and Herrerías (2001), Herrerías (2002), van Dorp and Kotz (2002a), (2002b) and (2003), García and García (2003) and Herrerías, García and Cruz (2003)). Likewise, the valuation method of the two distributions (VMTD) have been studied and applied, under uncertainty, to approach the market value of an asset from a quality index (see, e.g. García, Cruz and Andújar (1999), García, Trinidad and Gómez (1999), Cruz, García and García (2002), García, Cruz and García (2002), García, Cruz and Rosado (2002), Herrerías (2002), García and García (2003) and García, Herrerías and García (2003)); in this way, the valuation method based on the two survival functions (VMTS) have been also used to find the market value from a quality index (see, e.g. Callejón, Franco, Herrerías and Vivo (2005), Franco, Callejón, Herrerías and Vivo (2005) and Franco, Herrerías, Vivo and Callejón (2005)).

Unfortunately, the valuation methods based on two probability models present some disadvantages when it is considered a greater information by more than a quality index of this asset, such as loss or profit with respect to the assessments from each component of the quality index. In order to reduce loss of the assessments in risk analysis, it is useful to consider probability models weighing the distinct components of the quality index. Therefore, these procedures allow to correct and fit the market value of an asset. In particular, seeking to reduce the depreciation (appreciation) underwent in the market by the VMTD (VMTS) when more than one quality index is made available, it is common the use of weighted probability models based on the marginal distribution (survival) functions of the multidimensional quality index in both cases, independence and dependence among its components.

Therefore, these weighting procedures, marginal distribution or survival functions, in both independende and dependence cases, require to determine the weights, i.e., the coefficients α and p of the weighted probability models. Herrerías (2002) and García and García (2003) analyze three techniques to calculate these weights, and consequently, to generate weighted probability models:

1. subjective (supplied by an expert judgment).
2. modes (relationship between the modal values).
3. econometrics (fit by the linear models).

Within this framework, under independence of the components of the quality index, Herrerías (2002) summarizes the use of the three procedures, with their advantages and disadvantages, respectively. However, under dependence between the components of the quality index, he comments that these techniques are reduced to two: subjective and econometrics. Nevertheless, we have not found any motive to discard the mode technique under dependence, except the same ones and well-known in the case of independence.

Remark that, in the subjective technique, the expert (appraiser) supplies the information about the weights of the indexes, and so, one might point out the inconvenients of its subjectivity, see Herrerías (2002). Therefore, we analyze the remaining tools.

First, the weighting techniques to generate the weights are introduced, under both independence and dependence between the components of the quality index. Secondly, we expand these weighting tools by the survival functions, which allows us to generate other weighted probability models. Likewise, we discuss an alternative tool to determine the weights of the components of the quality index, modal mean technique, based on the modal values of its marginal distributions functions, which increases the possible weighted probability models to approach the market value of the asset. Finally, we give an application of these weighting procedures to generate weighted probability models in one example of land pricing, and thus we obtain the assessments of the land property according to each weighted probability model.

2. Techniques to generate weighted models valuation

In this section, we discuss the mode and econometric techniques, which were introduced by the marginal distribution functions. Subsequently, we analyze both methods from an alternative viewpoint, using the marginal survival functions.

2.1. *Econometric technique*

Let us see now, the econometric technique to generate the weights of the components of the quality index, in each case, independence and dependence.

2.1.1. *Econometric technique by distribution functions*

The generation of the weights by the econometric technique, when the components are independent, is based on the estimation of the following regression model

$$\log F_V(v_t) = \alpha \log F_1(i_{1t}) + \beta \log F_2(i_{2t}) + u_t, \; t = 1,\dots,n$$

where F_V is the distribution function of the market value V of the asset, F_i is the marginal distribution function of the i th component of the quality index, with $i = 1, 2$, and $F_I(i_1, i_2)$ its joint distribution function. So, the estimation $\hat{\alpha}$

of the coefficient of the first component might be calculated by restricted least squares, i.e., for least square involving the restriction $\alpha + \beta = 1$.

On the other hand, under dependence, the weights of the components using the econometric technique, is determined by the next regression model

$$F_V(v_t) = pF_1(i_{1t}) + qF_2(i_{2t}) + u_t,\ t = 1,\dots,n$$

and therefore, taking into account the restriction $p + q = 1$, the weight p of the first component is estimated by restricted least squares.

2.1.2. *Econometric technique by survival functions*

In this item, we propose a different viewpoint in the econometric technique, by the use of the survival function instead of the distribution function in the above regression models.

Thus, assuming independence between the components of the quality index, the estimation $\hat{\alpha}$ can be obtained through restricted least squares in the following regression model

$$\log S_V(v_t) = \alpha \log S_1(i_{1t}) + \beta \log S_2(i_{2t}) + u_t,\ t = 1,\dots,n$$

Likewise, when the components of the quality index are dependent, the estimation $\hat{p}$ might be calculated by restricted least squares in the next regression model

$$S_V(v_t) = pS_1(i_{1t}) + qS_2(i_{2t}) + u_t,\ t = 1,\dots,n$$

where $S_V = 1 - F_V$ is the survival function of the market value of the asset, and $S_i = 1 - F_i$ is the marginal survival function of the quality index, for $i = 1, 2$.

However, the advantage of the valuation methods of the two functions (VMTD and VMTS) against other appraisal methods is the practical utility of this methodology when only small samples are available and the weakness of the other ones in the absence of data. It reduces the usefulness of the above regression models, and consequently, for the estimation of the weights by the previous econometric techniques, which attempt to improve the assessments by the best fit among the distributions or survivals. Besides, one can note the addition of errors, in the estimation of the weights and the fit of the probability models (market value and quality index).

2.2. *Mode technique*

In this subsection, we analyze the generation of the weighted models by the mode technique, which is based on the equality of the modal values between the market value and the quality index through their probability models.

First, the mode technique allows us determine the weights by the following relationship between the distribution functions

$$F_V(m) = F_{WD}(m_1, m_2) \tag{1}$$

where F_{WD} is the distribution function of the weighted model from the marginal distribution functions of the quality index.

2.2.1. *Mode technique by distribution functions*

Let us see now the behaviour of the mode technique when both components of the quality index are independent, and secondly, when then components are dependent.

On the one hand, under independence of the components of the quality index, from Eq. (1) the weight of the first component is given by

$$F_V(m) = F_1^{\alpha}(m_1)F_2^{1-\alpha}(m_2)$$

or equivalently,

$$\frac{F_V(m)}{F_2(m_2)} = \left(\frac{F_1(m_1)}{F_2(m_2)}\right)^{\alpha} \tag{2}$$

where $0 < \alpha < 1$ represents the weight of the first component of the quality index.

If $F_1(m_1) = F_2(m_2)$, then Eq. (2) only makes sense when $F_V(m) = F_1(m_1) = F_2(m_2)$, which is a strong restriction on the modal market value of the asset. In that case, α might be any point in $(0,1)$.

If $F_1(m_1) \neq F_2(m_2)$, then the weight of the first component holds

$$\alpha = \frac{\log F_V(m) - \log F_2(m_2)}{\log F_1(m_1) - \log F_2(m_2)}$$

wherein we point out, the following contradictory situations:

1. When $F_V(m) < F_2(m_2) < F_1(m_1)$ then $\alpha < 0$
2. When $F_V(m) < F_1(m_1) < F_2(m_2)$ then $\alpha > 1$

Therefore, in order to that $\alpha \in (0,1)$, it is necessary to impose the following restriction on the modal market value of the asset

$$F_V(m) \in [F_i(m_i), F_j(m_j)] \text{ with } i \neq j \in \{1,2\} \text{ such that } F_i(m_i) < F_j(m_j) \tag{3}$$

Hence, we conclude with the following disadvantages of the mode technique by marginal distribution functions under independence:

1. The modal valuation has not to correspond with the modal quality index (see Ballestero and Rodríguez (1999) and Herrerías (2002)).
2. It is required strong restrictions on the modal value of the distribution function of the market value in order to generate feasible weights.

On the other hand, when the two components of the quality index are dependent, from Eq. (1), the mode technique allows us to get the weights by

$$F_V(m) = pF_1(m_1) + (1-p)F_2(m_2)$$

or equivalently,

$$F_V(m) - F_2(m_2) = p\left(F_1(m_1) - F_2(m_2)\right) \tag{4}$$

where $0 < p < 1$ represents the weight of the fisrt component of the quality index.

If $F_1(m_1) = F_2(m_2)$, then Eq. (4) only makes sense for $F_V(m) = F_1(m_1) = F_2(m_2)$, which is a strong restriction on the modal market value of the asset, and thus, p could be chosen in $(0,1)$.

If $F_1(m_1) \neq F_2(m_2)$, then the weights holds

$$p = \frac{F_V(m) - F_2(m_2)}{F_1(m_1) - F_2(m_2)}$$

wherein we point out, the following contradictory cases:

1. When $F_V(m) < F_2(m_2) < F_1(m_1)$ then $p < 0$
2. When $F_V(m) < F_1(m_1) < F_2(m_2)$ then $p > 1$

Therefore, in order to that $p \in (0,1)$, it is necessary to impose the restriction (3) on the modal market value.

Consequently, under dependence between the components of the quality index, the disadvantages of the mode technique by marginal distribution functions are the same as in the independent case.

2.2.2. *Mode technique by survival functions*

Similar to the econometric technique, we propose a different viewpoint in the mode technique based on the use of the survival functions instead of the distribution functions. So, the weights with the mode technique are determined by the following relationship between the survival functions

$$S_V(m) = S_{WS}(m_1, m_2) \tag{5}$$

where S_{WS} is the survival function of the weighted model from the marginal survival functions of the quality index.

So, let us see now, the behaviour of the mode technique from weighted models through the marginal survival functions of the bidimensional quality index, in both cases, independence and dependence of their components.

First, in the case of independence between the two components, from Eq. (5) the weight of the first component is given by

$$S_V(m) = S_1^{\alpha}(m_1) S_2^{1-\alpha}(m_2)$$

or equivalently,

$$\frac{S_V(m)}{S_2(m_2)} = \left(\frac{S_1(m_1)}{S_2(m_2)}\right)^{\alpha} \tag{6}$$

If $S_1(m_1) = S_2(m_2)$, then Eq. (6) only makes sense when $S_V(m) = S_1(m_1) = S_2(m_2)$, which is a strong restriction on the modal market value.

If $S_1(m_1) \neq S_2(m_2)$, then the weight holds

$$\alpha = \frac{\log S_V(m) - \log S_2(m_2)}{\log S_1(m_1) - \log S_2(m_2)}$$

wherein we remark the following contradictory situations:

1. When $S_V(m) > S_2(m_2) > S_1(m_1)$ then $\alpha < 0$
2. When $S_V(m) > S_1(m_1) > S_2(m_2)$ then $\alpha > 1$

Therefore, in order to that $\alpha \in (0,1)$, it is necessary to impose the following restriction on the modal market value of the asset

$$S_V(m) \in [S_i(m_i), S_j(m_j)] \text{ with } i \neq j \in \{1,2\} \text{ such that } S_i(m_i) < S_j(m_j) \tag{7}$$

Remark that the disadvantage of the mode techique by the weighting of the marginal survival functions is the strong restriction (7) on the modal value of the survival function of the market value in order to generate feasible weights.

On the other hand, under dependence between the components of quality index, from Eq. (5) the mode technique allows us to obtain the weights by

$$S_V(m) = pS_1(m_1) + (1-p)S_2(m_2)$$

or equivalently,

$$S_V(m) - S_2(m_2) = p\left(S_1(m_1) - S_2(m_2)\right) \tag{8}$$

where $0 < p < 1$ is the weight of the first component of the quality index.

If $S_1(m_1) = S_2(m_2)$, then Eq. (8) only makes sense for $S_V(m) = S_1(m_1) = S_2(m_2)$, which is a strong restriction on the modal market value.

If $S_1(m_1) \neq S_2(m_2)$, then the weights holds

$$p = \frac{S_V(m) - S_2(m_2)}{S_1(m_1) - S_2(m_2)}$$

wherein we remark the following contradictory cases:

1. When $S_V(m) > S_2(m_2) > S_1(m_1)$ then $p < 0$
2. When $S_V(m) > S_1(m_1) > S_2(m_2)$ then $p > 1$

Therefore, in order to that $p \in (0,1)$, it is necessary to impose the restriction (7) on the modal market value. So, the mode technique by marginal survival functions has the same disadvantage the mode in both cases, independence and dependence between of the components.

Besides, note that $S_1(m_1) = S_2(m_2)$ if and only if $F_1(m_1) = F_2(m_2)$, and the restrictions (3) and (7) are equivalent. Consequently, we have found the same disadvantages of this technique by both weightings, the marginal survival functions and the marginal distribution functions.

3. New technique to generate weighted models

In this section, we discuss a new technique to find the weights of the components, avoiding the disadvantages of the former tools: the subjectivity, the weakness of the econometric methods and the restrictions on the modal market value of the mode technique. For that, we consider a method based on the marginals of the probability model of the quality index and the weighted model from them, to generate the weights.

3.1. *Modal mean technique by distribution functions*

In order to generate weighted probability models by the marginal distribution functions, which reduce the depreciation of the VMTD with respect to the assessments from each component, the modal mean technique is based on the modal values of the quality index.

Remark that for any weight of the first component, $0 < \alpha < 1$ or $0 < p < 1$ under independence or dependence, respectively, the weighted model is bound by the marginal distribution functions

$$\inf\{F_1(i_1), F_2(i_2)\} \leq F_{WD}(i_1, i_2) \leq \sup\{F_1(i_1), F_2(i_2)\}$$

for all quality index (i_1, i_2). In particular, for the modal quality index (m_1, m_2), the following inequalities hold

$$\inf\{F_1(m_1), F_2(m_2)\} \leq F_{WD}(m_1, m_2) \leq \sup\{F_1(m_1), F_2(m_2)\}$$

In this setting, we propose the modal mean technique to generate the weighted model with parameter (α or p) such that the distance among these three values is minimized, i.e., the modal value of the weighted distribution function and the two modal values of the marginal distribution functions, which is given by

$$F_{WD}(m_1, m_2) = \frac{F_1(m_1) + F_2(m_2)}{2} \tag{9}$$

Note that the modal mean technique generates a weighted model using the available information by these quality indexes; so it is not influenced by the market value, and consequently, this procedure does not require any restriction on the modal market value.

Firstly, when the components of the quality index are independent, from Eq. (9) the weight α provides by the modal mean method is given by

$$F_1^{\alpha}(m_1)F_2^{1-\alpha}(m_2) = \frac{F_1(m_1)+F_2(m_2)}{2}$$

or equivalently,

$$\left(\frac{F_1(m_1)}{F_2(m_2)}\right)^{\alpha} = \frac{F_1(m_1)+F_2(m_2)}{2F_2(m_2)}$$

where $0<\alpha<1$ represents the weight of the first component of the quality index.

If $F_1(m_1)=F_2(m_2)$, then α can take any value in $(0,1)$.

If $F_1(m_1)\neq F_2(m_2)$, then the weight holds

$$\alpha = \frac{\log\left(\frac{F_1(m_1)+F_2(m_2)}{2}\right) - \log F_2(m_2)}{\log F_1(m_1) - \log F_2(m_2)} \in (0,1)$$

On the other hand, when the components of the quality index are dependent, from Eq. (9) we have the next equation

$$pF_1(m_1)+(1-p)F_2(m_2) = \frac{F_1(m_1)+F_2(m_2)}{2}$$

or equivalently,

$$p\left(F_1(m_1)-F_2(m_2)\right) = \frac{F_1(m_1)-F_2(m_2)}{2}$$

where $0<p<1$ represents the weight of the first component of the quality index.

If $F_1(m_1)=F_2(m_2)$, then p may be any point in $(0,1)$.

If $F_1(m_1)\neq F_2(m_2)$, then $p=1/2$.

Remark that under dependence between the components, the modal mean technique provides the same weight for each component of the quality index, which exhibits the coherence of the same one, since the dependent structure between the components includes the predominance and importance of one over the other, and therefore it will be contradictory the assignment of the different coefficients in the weighted model.

3.2. *Modal mean technique by survival functions*

Taking into account that for any weight of the first component, $0 < \alpha < 1$ or $0 < p < 1$ under independence or dependence, respectively, the weighted model is bound by the marginal survival functions

$$\inf\{S_1(i_1), S_2(i_2)\} \leq S_{WS}(i_1, i_2) \leq \sup\{S_1(i_1), S_2(i_2)\}$$

for all quality index (i_1, i_2), where the survival function S_{WS} of this weighted model allows to reduce the appreciation of the VMTS with respect to the assessments from each component.

In particular, for the modal quality index (m_1, m_2), we have

$$\inf\{S_1(m_1), S_2(m_2)\} \leq S_{WS}(m_1, m_2) \leq \sup\{S_1(m_1), S_2(m_2)\}$$

and therefore, the weighted model with coefficient (α or p), using the modal mean technique, is defined by the distance minimum among the modal value of the weighted model and the two modal values of the marginal survival functions,

$$S_{WS}(m_1, m_2) = \frac{S_1(m_1) + S_2(m_2)}{2} \tag{10}$$

Remark that this relationship to generate a weighted model using only the quality index, just like the modal mean method by distribution functions.

In the first place, when the components of the quality index are independent, from Eq. (10) the weight α of the first component is determined by

$$S_1^{\alpha}(m_1) S_2^{1-\alpha}(m_2) = \frac{S_1(m_1) + S_2(m_2)}{2}$$

or equivalently,

$$\left(\frac{S_1(m_1)}{S_2(m_2)}\right)^{\alpha} = \frac{S_1(m_1) + S_2(m_2)}{2S_2(m_2)}$$

If $F_1(m_1) = F_2(m_2)$, then α can be any value in $(0,1)$.

If $F_1(m_1) \neq F_2(m_2)$, then the weight holds

$$\alpha = \frac{\log\left(1 - \dfrac{F_1(m_1) + F_2(m_2)}{2}\right) - \log(1 - F_2(m_2))}{\log(1 - F_1(m_1)) - \log(1 - F_2(m_2))} \in (0,1)$$

In the second place, when the components of the quality index are dependent, from Eq. (10) the weight p of the first component is given by

$$pS_1(m_1) + (1-p)S_2(m_2) = \frac{S_1(m_1) + S_2(m_2)}{2}$$

or equivalently,

$$p\left(S_1(m_1)-S_2(m_2)\right)=\frac{S_1(m_1)-S_2(m_2)}{2}$$

If $F_1(m_1)=F_2(m_2)$, then p can be chosen in $(0,1)$.

If $F_1(m_1)\neq F_2(m_2)$, then $p=1/2$.

Consequently, under dependence between the components and using their marginal survival functions, the modal mean technique provides the same weights for each component of the quality index, just like when one uses the marginal distribution functions, and therefore it will be contradictory the assignment of the different coefficients in the weighted model, because its dependence includes the prevalence of one over the other.

4. Practical application

In this section, we expose a practical application of the weighted probability models, by these techniques, in one example of land pricing. In particular, we consider the transactions of agricultural propierties in *Tierras de Campos* and *Centro* regions (Valladolid, Spain) given in Alonso and Lozano (1985) and García and García (2003). The quality indexes used to explain the market values (€) are the income per hectare (*€/Ha*) and the inverse distance to Valladolid (1/*km*).

Table 1 displays data of the minimum (pessimistic), maximum (optimistc) and modal (most likely) values for each variable; where the objective is to appraise an agricultural property whose income per hectare is 194.31 and location is *km* from Valladolid.

Table 1. Transactions of agricultural plots

	Pessimistic	Optimistic	Most likely
V=Market value (€/Ha)	1502.53	3005.06	1953.29
I_1=Income (*€/Ha*)	120.20	300.51	195.33
I_2=Inverse distance (1/*km*)	1/70	1/10	1/50

Remark that the independence between both components of the bidimensional quality index is assumed in this application by García and García (2003), as well as their triangular distributions, and to reduce the depreciation of the VMTD with respect to the appraisals obtained for each component, the weight of the first component, $\alpha=0.75$, is provided by an expert judgment. Besides, we will consider that the market value of an agricultural plot follows a triangular or trapezoidal model, which are a sample of the different models that might be considered.

Thus, in Tables 2 and 3, the weights of the first component have been determined by the marginal distribution functions, and taking into account the same ones for a better comparison in both cases, triangular and trapezoidal market value when the weighting technique is econometric or mode, since the other procedures are not influenced by the distribution function of the market value.

In particular, Table 2 displays the assessments of the property through both methods, VMTD and VMTS, when the weighted probability models are based on the marginal distribution functions

Table 2. Weighted model of the marginal distribution functions (F_{WD})

Weighting Technique	α	F_V	v_{DWD}	v_{SWD}
Subjective	0.75	Triangular	2054.33	2655.11
Subjective	0.75	Trapezoidal	2113.79	2681.07
Econometric	0.615456	Triangular	2064.94	2609.91
Econometric	0.615456	Trapezoidal	2125.23	2639.22
Mode	0.82074	Triangular	2048.92	2695.77
Mode	0.82074	Trapezoidal	2107.90	2718.71
Modal Mean	0.702754	Triangular	2058.01	2635.08
Modal Mean	0.702754	Trapezoidal	2117.77	2662.52

and Table 3 includes the appraisals of the weighted probability models based on the marginal survival functions

Table 3. Weighted model of the marginal survival functions (S_{WD})

Weighting Technique	α	F_V	v_{DWS}	v_{SWS}
Subjective	0.75	Triangular	1689.98	2057.37
Subjective	0.75	Trapezoidal	1707.88	2117.09
Econometric	0.615456	Triangular	1711.83	2068.83
Econometric	0.615456	Trapezoidal	1731.80	2129.40
Mode	0.82074	Triangular	1669.16	2051.29
Mode	0.82074	Trapezoidal	1685.07	2110.49
Modal Mean	0.702754	Triangular	1699.95	2061.41
Modal Mean	0.702754	Trapezoidal	1718.79	2121.44

Analogously, in Tables 4 and 5, the weights of the first component are generated from the marginal survival functions.

In particular, Table 4 shows the valuations when the weighted probability models are based on the marginal distribution functions by both methods, VMTD and VMTS

Table 4. Weighted model of the marginal distribution functions (F_{WD})

Weighting Technique	α	F_V	v_{DWS}	v_{SWS}
Econometric	0.671763	Triangular	2060.45	2624.50
Econometric	0.671763	Trapezoidal	2120.40	2652.73
Mode	0.612085	Triangular	2065.21	2609.22
Mode	0.612085	Trapezoidal	2125.52	2638.58
Modal Mean	0.441782	Triangular	2079.29	2598.49
Modal Mean	0.441782	Trapezoidal	2140.51	2628.65

and Table 5 displays the assessments using the weighted probability models through the marginal survival functions

Table 5. Weighted model of the marginal survival functions (S_{WD})

Weighting Technique	α	F_V	v_{DWS}	v_{SWS}
Econometric	0.671763	Triangular	1705.06	2064.05
Econometric	0.671763	Trapezoidal	1724.39	2124.28
Mode	0.612085	Triangular	1712.13	2069.12
Mode	0.612085	Trapezoidal	1732.14	2129.70
Modal Mean	0.441782	Triangular	1714.64	2083.42
Modal Mean	0.441782	Trapezoidal	1734.89	2144.86

Remark that in all cases, the VMTS proposes appraisals greater than the VMTD.

Besides, we show some graphics on the behaviour of both VMTS and VMTD from the weighted models, in which the market value of an agricultural land follows a triangular model and the quality index has independent and triangular components.

In the first place, from marginal distribution functions, $\hat{\alpha} = 0.82074$, 0.615456 and 0.702754 by the mode, econometric and modal mean techniques, respectively, and the valuation obtained from these procedures will be marked by "m", "e" and "mm", respectively. So, Figures 1 and 2 describe the assessments by the weighted models of the marginal distribution and survival functions, respectively, where $V1$ and $V2$ are the valuations from each component of the quality index.

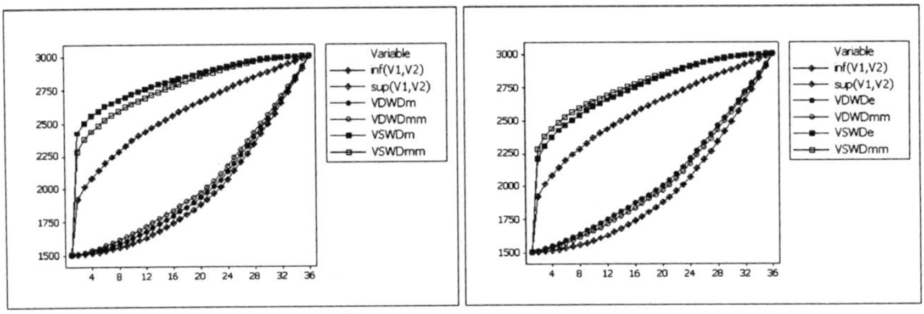

Figure 1. Weighted models of the marginal distribution functions

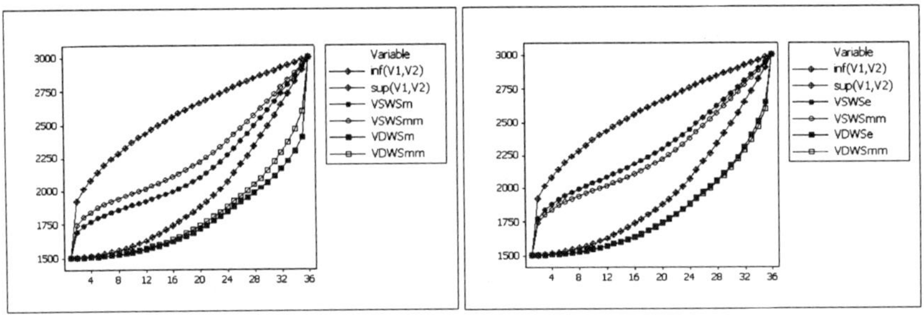

Figure 2. Weighted models of the marginal survival functions

Similarly, from marginal survival functions, $\hat{\alpha} = 0.612085$, 0.671763 and 0.441782 by the mode, econometric and modal mean methods. Thus, Figures 3 and 4 depict the appraisals by the weighted probability models of the marginal distribution and survival functions, respectively.

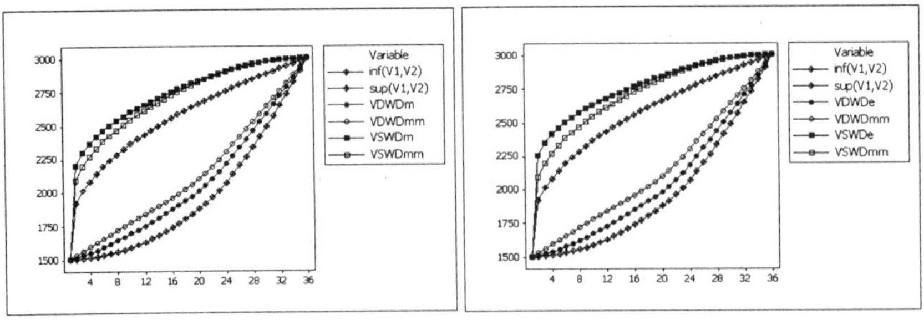

Figure 3. Weighted models of the marginal distribution functions

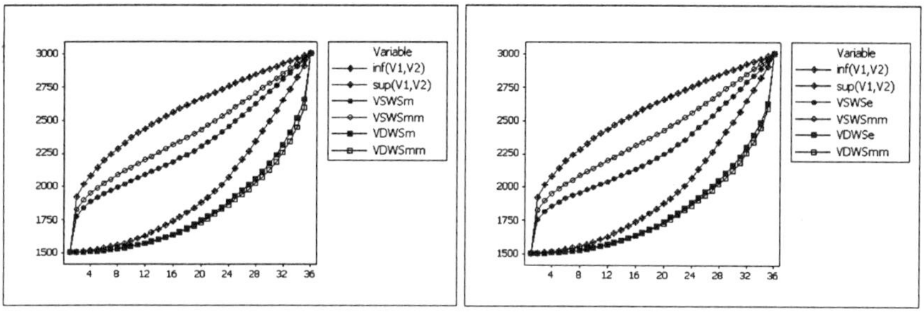

Figure 4. Weighted models of the marginal survival functions

5. Comments and conclusions

Finally, we give some comments and we point out the main conclusions of this work.

In the subjective technique, the expert judgment (appraiser) supplies the information about the weights of the quality indexes, and its subjectivity was commented by Herrerías (2002).

The main advantage of the valuation methods of the two functions (VMTD and VMTS), against other appraisal methods, is the usefulness in absence of data; in that situation, it is known the weakness of the regression models, and consequently, the estimation of the weights by the econometric technique in the generation of weighted models to correct and fit the assessments.

Likewise, the mode technique also has some disadvantages, in general, the modal valuation has not to correspond with the modal quality index (see Ballestero and Rodríguez (1999) and Herrerías (2002)). Furthermore, it is required a strong restriction on the modal market value in order to generate feasible weights, in both cases, independence and dependence between the components of the quality index.

Finally, the modal mean technique allows to generate weighted models avoiding the subjectivity of the expert judgment, the weakness of the econometric methods and the restrictions on the modal market value. Besides, the modal mean procedure provides feasible weights for the components of the quality index.

References

1. Alonso, R. and Lozano, J. (1985). El método de las dos funciones de distribución: Una aplicación a la valoración de fincas agrícolas en las

comarcas Centro y Tierra de Campos (Valladolid). Anales del INIA: Economía, 9, 295–325.
2. Ballestero, E. and Rodríguez, J.A. (1999). El precio de los inmuebles urbanos. CIE Inversiones, Ed. DOSSAT 2000, Madrid.
3. Callejón, J., Franco, M., Herrerías, R. and Vivo, J.M. (2005). El método de valoración de las dos funciones de supervivencia como metodología alternativa al de las dos funciones de distribución. In XIX Reunión ASEPELT-ESPAÑA, Badajoz.
4. Callejón, J., Pérez, E. and Ramos, A. (1996). La distribución trapezoidal como modelo probabilístico para la metodología PERT. In X Reunión de ASEPELT-ESPAÑA, Albacete. Content in Programación, Selección y Control de Proyectos en ambiente de incertidumbre. R. Herrerías (ed.). Universidad de Granada, 2001, 167–177.
5. Cruz, S., García, C.B. and García, J. (2002). Statistical test for the method of the two distribution functions. An application in finance. In VI Congreso de Matemática Financiera y Actuarial and 5th Italian-Spanish Conference in Financial Mathematics, Valencia.
6. Franco, M., Callejón, J., Herrerías, R. and Vivo, J.M. (2005). Procedimiento para reducir la depreciación del valor de mercado del método de valoración de las dos funciones de distribución: Funciones de supervivencia y máximo. In VI Seminario de ASEPELT sobre Análisis, Selección, Valoración, Control y Eficiencia de Proyectos, Murcia.
7. Franco, M., Herrerías, R., Vivo, J.M. and Callejón, J. (2005). Valuation method of the two survival functions as a proxy methodology in risk analysis. In CIMMA2005: International Mediterranean Congress of Mathematics Almería 2005, Almería.
8. García, J., Cruz, S. and Andújar, A.S. (1999). Il metodo delle due funzioni di distribuzione: Il modello triangolare. Una revisione. Genio Rurale, 11, 3–8.
9. García, J., Cruz, S. and García, L.B. (2002). Generalización del método de las dos funciones de distribución (MTDF) a familias beta determinadas con lo tres valores habituales. In III Reunión Científica ASEPELT: Análisis, Selección, Control de Proyectos y Valoración, Murcia, 89–113.
10. García, J., Cruz, S. and Rosado, Y. (2002). Extensión multi-índice del método beta en valoración agraria. Economía Agraria y Recursos Naturales, 2, 3–26.
11. García, J. and García, L.B. (2003). Teoría General de Valoración. Método de las dos funciones de distribución. Ed. Fundación Unicaja, Málaga.
12. García, J., Herrerías, R. and García, L.B. (2003). Valoración agraria: Contrastes estadísticos para índices y distribuciones en el método de las dos funciones de distribución. Revista Española de Estudios Agrosociales y Pesqueros, 199, 93–118.

13. García, J., Trinidad, J.E. and Gómez, J. (1999). El método de las dos funciones de distribución: la versión trapezoidal. Revista Española de Estudios Agrosociales y Pesqueros, 185, 57–80.
14. Herrerías, J.M. (2002). Avances en la Teoría General de Valoración en Ambiente de Incertidumbre. Tesis Doctoral. Universidad de Granada.
15. Herrerías, R., García, J. and Cruz, S. (2003). A note on the reasonableness of PERT hypotheses. Operations Research Letters, 31, 60–62.
16. Herrerías, R., García, J., Cruz, S. and Herrerías, J.M. (2001). Il modello probabilistico trapezoidale nel metodo delle due distribuzione della teoria generale de valutazioni. Genio Rurale. Rivista di Scicienze Ambientali, LXIV, 3–9.
17. Johnson, D. (1997). The triangular distribution as a proxy for the beta distribution in risk analysis. Journal of the Royal Statistical Society, Ser. D, 46, 387–398.
18. Johnson, N.L. and Kotz, S. (1999). Non-smooth sailing or triangular distributions revisited after some 50 years. Journal of the Royal Statistical Society, Ser. D, 48, 179–187.
19. van Dorp, J.R. and Kotz, S. (2002a). The standard two sided power distribution and its properties: With applications in financial engineering. The American Statistician, 56, 90–99.
20. van Dorp, J.R. and Kotz, S. (2002b). A novel extension of the triangular distribution and its parameter estimation. Journal of the Royal Statistical Society, Ser. D, 51, 63–79.
21. van Dorp, J.R. and Kotz, S. (2003). Generalized trapezoidal distributions. Metrika, 58, 85–97.
22. Williams, T.M. (1992). Practical use of distributions in network analysis. Journal of the Operational Research Society, 43, 265–270.

Chapter 5

ON GENERATING AND CHARACTERIZING SOME DISCRETE AND CONTINUOUS DISTRIBUTIONS

M.A. FAJARDO-CALDERA

Dpto. de Economía Aplicada y Organización de Empresas
University of Extremadura
Camino de Elbas s/n, Badajoz, 06071, Spain

J. PÉREZ-MAYO

Dpto. de Economía Aplicada y Organización de Empresas
University of Extremadura
Camino de Elbas s/n, Badajoz, 06071, Spain

The main aim of this paper is to generate compound distributions, discrete or continuous, from Binomial conditional distributions by means of Bayesian techniques. Besides, the authors extend Kowar's paper (1975) by characterizing some discrete and continuous distributions, in the context of some well-known distributions, from the conditional distribution of a random variable (r.v.) and by the linear regression of the latter given the former.

1. Introduction

One of the main aims of Probability Calculus is to determine some theoretical distributions useful for modeling the random phenomena that appear in Experimental Sciences.

Many methods have been applied to generate or characterize discrete or continuous distributions of probability: change of variables, functional equations (differential or in differences), etc. These methods supply the theoretical issues needed to describe a random phenomenon and to obtain the explicit probability law.

In the direct method, some distributions are obtained by the expression of a mathematical model, which is the abstraction of a random experiment. An example of this method is the theory of combinatory numbers to directly get the probabilities that correspond to each value of a random variable. From this theory some important and well-known distributions as the binomial, hypergeometric, geometric or negative binomial ones, are generated.

Sometimes, while trying to establish the model, one must solve an equation (functional, differential, in differences) to explicitly obtain the probability law. For example, the equation in differences obtained when one establish the probability of getting r successes in n independent tests, by assuming that the probability of success varies in each test. The equation appears in the generalization of the repeated Bernoulli's tests, having the Binomial distribution as a particular case.

It is also possible to start from a differential equation, being the Poison distribution the most known. Systems of differential equations have been proposed in the literature. The most important one is the well-known Pearson's system of curves, that is a generalization of the differential equation generated from the Normal distribution and whose solution contains many of the continuous distributions of probability as the Normal, Gamma, Beta, Exponential. Later, this system was studied by Elderton and Johnson (1969) and extended by Herrerías Pleguezuelo (1975) and Callejón (1995).

The discrete consideration of the Pearson's system is done by Ord (1972) and generalized by Fajardo (1985) and Rodriguez Avi (1993), whose most important consequence is the extended analysis of the family of discrete distributions of probability defined by the generalized hypergeometric series. This analysis was done by Dacey (1971) and later extended and generalized by Hermoso Gutierrez (1986) and Saez Castillo (2002).

An alternative method used in Statistics for generating distributions of probability is the use of functions of random variables, i.e. variables transformations. The most usual transformations are the sum, product or quotient of two variables.

Finally, it can not be forgotten another method of getting probabilistic distributions by means of limits. Two well-known examples are the conversion of a Binomial distribution into a Poisson or a Normal one.

Following the steps above, in this paper we try to generate probability distributions from compound distributions by means of the well-known Bayesian techniques and, on the other hand, to characterize discrete distributions from Binomial conditional distributions and linear regression.

2. The binomial model

In many statistical experiments, the observations are considered to be generated by a binomial distribution. This distribution describes discrete data, resulting from an experiment called Bernoulli's process in Jacob Bernoulli's honour.

Consider a population in which an event happens as the outcome of a Bernoulli trial with probability p. Thus, given $0 < p < 1$, the number of occurrences for k in r trials has the binomial distribution,

$$P[k \mid r, p] = \binom{r}{k} p^k (1-p)^{r-k}, k = 0,1,\ldots r \tag{1}$$

It is necessary to be careful in using the binomial distribution because the following conditions must be fulfilled: each trial only has two possible results, the probability of each trial remains fixed along the time and the trials are statistically independent. Specifically, the second and third conditions require that the probability of results in every trial remains fixed along the time and the trials or attempts in a Bernoulli process are statistically independent, that is, the result of a trial can not affect the result of any other trial.

3. Generating compound distributions

As it is commented in the section before, necessary conditions for using the binomial distribution, are not satisfied in the most of experiments because, in general, parameters *(r, p)* are usually random instead of fixed and, therefore, one needs to define a distribution of probability.

Compound distributions such as the compound Poisson and the compound negative binomial are used extensively in the theory of risk to model the distribution of the total claims incurred in a fixed period of time

Considered the former, one can take into account, theoretically, a random variable ξ depending on a parameter θ, which is also a random variable. The distribution of both variables is:

$$P[\xi = s, \theta = n] = P[\xi = s \mid \theta = n] P[\theta = n] \tag{2}$$

In many situations, the main interest lies in the *marginal distribution* of ξ to predict instead of the value of parameter θ. This marginal distribution is called in the statistical literature *compound distribution* and can be obtained:

$$P[\xi = s] = \int f[\Theta = \theta] P[\xi = s \mid \Theta = \theta] d\theta \tag{3}$$

if θ is continuous.

If θ is discrete and finite, then the compound distribution is called *mixture of distributions* and is given by

$$P[\xi - s] = \sum_n P[\xi = s \mid \Theta = \theta] P[\Theta = \theta] \tag{4}$$

4. A compound distribution when the parameters of a binomial distribution are variable (n) and fixed (p)

Assume the case of a random variable ξ with a binomial distribution with a parameter θ, being p fixed, as it is defined below:

$$P[\xi = s \mid \theta] = \binom{\theta}{s} p^s (1-p)^{\theta - s}, s = 0,1,\ldots,\theta \tag{5}$$

Assume that θ varies across the population according to a Poisson distribution with λ parameter

$$P[\theta = n] = \frac{\lambda^n}{n!} e^{-\lambda} \tag{6}$$

Then, from (8) and (9) one obtains the marginal distribution of ξ, a Poisson-Binomial compound distribution, given by:

$$P[\xi = s] = \sum_{n=0}^{\infty} \binom{\theta}{s} p^s (1-p)^{\theta - s} \frac{\lambda^n}{n!} e^{-\lambda} = \frac{e^{-\lambda p} (\lambda p)^s}{s!} \tag{7}$$

This distribution matches the Poisson distribution with parameter λp.

As an application, consider such a population that the probability of having a male newborn is the parameter p (fixed) and one wants to compute the probability of having s males in a family.

If the probability for a newborn to be male is $p = 0.51$ and the probability of having s males in a family of n children is given by:

$$\binom{n}{s} 0.51^s (1-0.51)^{n-s}$$

Then, the probability for any family to have s male children will be:

$$\frac{(0.51\lambda)^s}{s!} e^{-0.51\lambda}$$

by assuming that $\frac{\lambda^n}{n!} e^{-\lambda}$ is the probability of having n children.

5. A compound distribution when the parameters of a binomial distribution are variable (p) and fixed (n)

Consider that the distribution of probability of the random variable ξ given in (1) depends on the parameter p, being r fixed. Therefore, one can generate a mixture of distributions if consider the following theoretical situation:

$$P[\xi = k \mid r, \theta] = \binom{r}{k} \theta^k (1-\theta)^{r-k}, k = 0,1,\ldots,r \tag{8}$$

Suppose that θ varies across the population according to a Beta distribution,

$$f(\theta \mid \alpha, \beta) = \frac{\theta^{\alpha-1}(1-\theta)^{\beta-1}}{Be(\alpha, \beta)}, \alpha > 0, \beta > 0, 0 < \theta > 1 \quad (9)$$

where $B(\alpha,\beta)$ is the complete beta function. Since θ is not observable, the probability distribution of k in r trials, given α and β, for a randomly chosen member is the following simple mixture model

$$P[\xi = k \mid r, \alpha, \beta] = \int P[\xi = k \mid r, \theta] f(\theta, \alpha, \beta) d\theta \quad (10)$$

By using (8) and (9), the probability distribution of ξ is defined as

$$P[\xi = k \mid r, \alpha, \beta] = \binom{r}{k} \frac{Be(\alpha + k, \beta + r - k)}{Be(\alpha, \beta)}, k = 0, 1, \ldots, r \quad (11)$$

This compound distribution is the beta-binomial model.

As an application of this beta-binomial model, consider that one wants determine the probability of k, randomly selected, elements in a population have influenza, knowing that the initial distribution of the proportion θ of elements with influenza is the distribution $Be\ (\theta | 3, 12)$ and that a random sample of 20 elements contained five sick people. Then, considering the former, the probability for k randomly selected elements in a population have influenza comes is given by

$$P[\xi = k \mid 20, 3, 12] = \binom{20}{k} \frac{Be(3 + k, 12 + 20 - k)}{Be(3, 12)}$$

6. Poisson compound distributions

It is usual to find situations where the Binomial distribution is applied with a little p and a large n. Assuming $rn = \lambda$ constant and taking limits for the binomial distributions when $n \to \infty$, being p variable, one obtains a distribution known as Poisson's Law. This reasoning means that Poisson's distribution approximate the Binomial distribution rather well if n is large y p small.

Then, considering the former circumstances, the probabilistic distribution given by (1) becomes into a Poisson's distribution con this conditional density function given $\Theta = \theta$:

$$f(y / \theta) = \frac{\theta^y}{y!} e^{-\theta}, \text{ with } \theta > 0 \text{ and } y = 0, 1, 2, \ldots \quad (12)$$

Let Θ be a r.v. with density function $f(\theta)$ and assume that Θ follows a Gamma distribution, given by:

$$f(\theta)=\begin{cases}\dfrac{a^{u}}{\Gamma(u)}\theta^{u-1}e^{-a\theta}, \theta>0\\ 0, \theta\leq 0\end{cases}, \text{ with } u>0,\ a>0 \tag{13}$$

Then, the compound Gamma-Poisson distribution is given by:

$$P[Y=y]=\int_0^{\infty}\frac{\theta^{y}}{y!}e^{-\theta}\frac{a^{u}}{\Gamma(u)}\theta^{u-1}e^{-a\theta}d\theta=\frac{a^{u}}{\Gamma(u)}\int_0^{\infty}\frac{\theta^{y+u-1}}{y!}e^{-(a+1)\theta}d\theta=(-1)^{y}\binom{-u}{y}p^{y}q^{u} \tag{14}$$

That is known as negative binomial distribution of *(u,p)* parameters. This distribution, introduced by Greenwood and Yule, has been used to represent the industrial accidents. It is possible to say that the probability of y accidents is given by a Poisson distribution of θ parameter. However, this parameter changes for each worker and one can observes that, assuming a Gamma distribution, the observations follows a Poisson compound distribution.

7. On characterizing some discrete distributions by linear regression

In this section, we consider characterizations of discrete distributions, in the context of some well known bivariate distributions, by the conditional distribution of one random variable (r.v.) given another one, and the linear regression of the second r.v. on the first.

Let X be a discrete r.v. taking the values $0...m$, where m may be a positive integer or ∞. Let the conditional distribution of another r.v. Y given X be given by:

$$P[y=y\mid X=x]=\binom{x}{y}p^{x}q^{x-y} \tag{15}$$

with $0<p<1$ and $q=1-p$. Here p is independent of x.

Finally, let

$$E[X\mid Y=y]=ay+b \tag{16}$$

where a and b are constant.

The question is: What distributions X are characterized by (15) and (16)?

We have the following propositions that prove, among other things, that a and b in (16) must be positive.

Proposition 6.1

In the previous conditions with a finite E(X), we have the following: (A) $b>0$; (B) X is bounded if, and only if, $0<a<1$. If X is bounded, then $b=m(1-a)$; (C) $0<a<p^{-1}$.

Proof.

(A) Specifying (16) for $y = 0$, we have $E[X \mid Y = 0] = b$. On the other hand,

$$P[Y = y] = \sum_{x=y}^{m} P[X = x, Y = y] = \sum_{x=y}^{m} P[X = x] P[Y = y \mid X = x] = \sum_{x=y}^{m} \binom{x}{y} p^y q^{x-y} P[X = x] \tag{17}$$

For $y = 0$ and $y = m$, we have that:

$$P[Y = 0] = \sum_{x=0}^{m} q^x P[X = x] \tag{18}$$

and

$$P[Y = m] = \sum_{x=m}^{m} P[X = m] \binom{x}{m} p^x q^{x-m} = p^m P[X = m] \tag{19}$$

Then, since $X \geq 0$ we have that

$$0 \leq E[X \mid Y = 0] = \sum_{x=0}^{m} x P[X = x \mid Y = 0] = \sum_{x=0}^{m} \frac{x P[X = x] P[Y = 0 \mid X = x]}{P[Y = 0]} = \sum_{x=0}^{m} x q^x \frac{P[X = x]}{P[Y = 0]} = b$$

If $b = 0$, then $P[X = x] = 0$ for $x = 1, 2 \ldots\ldots m$ and hence X is degenerate at 0 which is a contradiction. Then $b > 0$.

(B) Let X be bounded, that is, let m a positive integer. Then, specifying (2) for $y = m$, we have

$$E[X \mid Y = m] = am + b \tag{20}$$

On the other hand,

$$E[X \mid Y = m] = \sum_{x=0}^{m} x P[X = x \mid y = m] = \sum_{x=m}^{m} x \frac{\binom{x}{m} p^m (1-p)^{x-m} P[X = x]}{P[Y = m]} = \frac{m p^m P[X = x]}{P[Y = m]} = m \tag{21}$$

Then, from (20) and (21), we have that $m = E[X \mid Y = m] = am + b$. Then, $b = m(1 - a)$ implies that $0 < a < 1$.

Now, assume that X is unbounded. Noting that Y also takes values 0, 1, 2..., and $Y \leq X$, we have from (16) that $y = 0, 1, 2\ldots$

$$y = \sum_{x \geq y} y P[X = x / Y = y] \leq \sum_{x \geq y} x P[X = x / Y = y] = E[X / Y = y] = ay + b \tag{22}$$

or $(1 - a)\, y \leq b$; which cannot hold for all the nonnegative integers unless $(1 - a) \leq 0$. This proves (B).

(C) By considering that

$$E[X \mid Y=y]=\sum_x xP[X=x \mid Y=y]=\sum_x x\frac{P[X=x,Y=y]}{P[Y=y]}=ay+b \quad (23)$$

and by summing both sides of (23), one can obtain that

$$\sum_x\sum_y xP[X=x,Y=y]=\sum_y (ay+b)P[Y=y] \quad (24)$$

From (24) we have

$$E[X]=aE[Y]+b \quad (25)$$

Besides,

$$E[X \mid Y=y]=\sum_y yP[Y=y \mid X=x]=\sum_y y\frac{P[X=X,Y=y]}{P[X=x]}=xp \quad (26)$$

and by summing both sides of equation (26), we have

$$E[Y]=\sum_x\sum_y xP[X=x,Y=y]=\sum_x xpP[X=x]=pE[X] \quad (27)$$

and, therefore, from equations (25) and (27), one can obtain that

$$E[X]=\frac{b}{(1-ap)}>0$$

from which

$$a>1/p$$

as we want to prove in (C).

Theorem 6.1

Assume that X is a discrete r.v. taking the values 0,..., *m*, where *m* may be a positive integer or ∞. Let E[X] finite. Let Y be another r.v. whose conditional distribution given X is given by (15). Then (16) hold for some constants *a*, *b* if, and only if, X has these distributions: binomial iff $0 < a < 1$, negative binomial and Poisson. Furthermore, X is binomial iff $0 < a < 1$, negative binomial iff $a < 1$, and Poisson iff $a = 1$.

Proof (⇒):

Let (16) hold for some constants *a* and *b*. Letting P[X = *x*], *x* = 0,..., *m*, and G (t) = E[t^X], the probability generating function (p.g.f) of X, we have

$$E[X/Y=y]=\sum_{x=y}^{m} xP[X=x/Y=y]=\sum_{x=y}^{m} x\frac{P[X=x]P[Y=y/X=x]}{P[Y=y]}$$
$$=\sum_{x=y}^{m} x\binom{x}{y}\frac{p^y q^{x-y}P[X=x]}{P[Y=y]}=ay+b \quad (28)$$

$y = 0, 1, \ldots m$. That is,

$$\sum_{x=y}^{m} x\binom{x}{y}\frac{p^{y}q^{x-y}P[X=x]}{P[Y=y]} = (ay+b)P[Y=y] \tag{29}$$

If we use the following expression, $x\binom{x}{y} = (y+1)\binom{x}{y+1} + y\binom{x}{y}$, we have,

$$(y+1)\frac{q}{p}\sum_{x=y}^{m}\binom{x}{y+1}p^{y+1}q^{x-(y+1)}P[X=x] + y\sum_{x=y}^{m}\binom{x}{y}p^{y}q^{x-y}P[X=x] = (ay+b)P[Y=y] \tag{30}$$

Then by (30), we have.

$$(y+1)\frac{q}{p}P[Y=y+1] = (a-1)yP[Y=y] + bP[Y=y] \tag{31}$$

By multiplying both sides of (31) by t and summing over $y = 0, 1 \ldots, m$, we have the differential equation

$$q/_{p}\,H'(t) = (a-1)tH'(t) + bH(t) \tag{32}$$

for the p.g.f. $H(t)$ of Y. However, it is known that the p.g.f. of Y is given by

$$H(t) = \sum_{y=0}^{x} t^{y}P[Y=y] = \sum_{y=0}^{x} t^{y}\sum_{x=y}^{m}\binom{x}{y}p^{y}q^{x-y}P[X=x] = \sum_{y=0}^{x}\sum_{x=y}^{m}\binom{x}{y}(tp)^{y}q^{x-y}P[X=x]$$

$$= \sum_{x=y}^{m}\left\{\sum_{y=0}^{x}\binom{x}{y}(tp)^{y}q^{x-y}\right\}P[X=x] = \sum_{x=y}^{m}(tp+q)^{x}P[X=x] = G(tp+q) \tag{33}$$

From (32) and (33), we get the differential equation

$$qG'(t_1) = (a-1)(t_1-q)G'(t_1) + bG(t_1) \tag{34}$$

for the p.g.f. $G(t_1)$ of X, (remember that $t_1 = pt + q$). To solve (34) for $G(t_1)$, we consider two cases: (i) $a = 1$ and (ii) $a \neq 1$.

When $a = 1$, $qG'(t_1) = bG(t_1)$, then

$$G(t_1) = ce^{bt_1/q} \tag{35}$$

with $c = e^{-b/q}$.

Thus, $G(t) = e^{b(t-1)/q}$, showing thereby that X is Poisson with $\lambda = b/q$ $(b > 0)$, [remember that $G(t) = e^{\lambda(t-1)}$, with $\lambda > 0$, is the p.g.f. of Poisson distribution]

When $a \neq 1$, then the equation (34) can be written as

$$G'(t_1)\Big/G(t_1) = b\Big/\{q-(a-1)(t_1-q)\}$$

whose solutions are

$$G(t_1) = c\{q-(a-1)(t_1-q)\}^{-\nu}, \quad \nu = b/(a-1), \text{ and } c = (1-ap)^{\nu} \tag{36}$$

Thus,

$$G(t_1) = \left\{ aq\Big/(1-ap) + (1-a)t_1\Big/(1-ap) \right\}^{-\nu} \tag{37}$$

Now, let $0 < a < 1$. Then, proposition 1.C. (B), shows that $b = m(1 - a)$. Thus, considering the equation (37), we have

$$G(t_1) = \left\{ aq\Big/(1-ap) + (1-a)t_1\Big/(1-ap) \right\}^{m} \tag{38}$$

From (38) it follows that X has a binomial distribution with parameters (m,α), where $\alpha = (1 - a)/(1 - ap)$. [Remember that $G(t) = (pt + q)^n$ is the p.g.f. of binomial distribution]

Finally, assume $a > 1$. Considering the equation (37), it follows that X has a negative binomial distribution with parameters (ν,α), where $\alpha = (1 - ap)/aq$. For proposition 1.C. (C) we have $ap < 1$, then α is indeed positive. [Remember that $G(t) = (p/1 - qt)^n$ is the p.g.f. of negative binomial distribution]

This proves the "only if" part of the theorem.

⇐) To prove the "if" part of the theorem, we consider the following:

Suppose that X is Poisson and $Y/X = x$ is given by (1). Then, $E[X|Y = y] = ay + b$, where a and b are constant.

Let be

$$P[X=x,Y=y] = P[X=x]P[Y=y \mid X=x] = e^{-\lambda}\lambda^{x}\Big/x! \binom{x}{y} p^{y} q^{x-y}$$
$$= e^{-\lambda}(\lambda p)^{y}(\lambda q)^{x-y}\Big/y!(x-y)! \tag{39}$$

the joint distribution of (X,Y) bivariate random variables (b.r.v).

Then,

$$P[Y=y]=\sum_{x=0}P[X=x,Y=y]=\sum_{x=y}e^{-\lambda}(\lambda p)^{y}(\lambda q)^{x-y}\Big/y!(x-y)!$$
$$=e^{-\lambda}(\lambda p)^{y}/y!\sum_{x=y}\frac{(\lambda q)^{x-y}}{(x-y)!}=e^{-\lambda}e^{\lambda q}(\lambda p)^{y}/y!=\frac{e^{-\lambda p}(\lambda p)^{y}}{y!} \tag{40}$$

Therefore, Y has a Poisson distribution with parameter λp.

From (39) and (40), we have that:

$$P[X=x\mid Y=y]=\frac{P[X=x,Y=y]}{P[Y=y]}=\frac{e^{-\lambda}(\lambda p)^{y}(\lambda q)^{x-y}}{y!(x-y)!}\Bigg/\frac{e^{-\lambda p}(\lambda p)^{y}}{y!}=\frac{e^{-\lambda q}(\lambda q)^{(x-y)}}{(x-y)!} \tag{41}$$

Therefore, X|Y $= y$ follows a Poisson distribution with parameter λq. The E[X/Y $= y$] is the following:

$$E[X=x\mid Y=y]=\sum_{x\geq y}x\frac{e^{-\lambda q}(\lambda q)^{(x-y)}}{(x-y)!}=\sum_{x\geq y}(x-y+y)\frac{e^{-\lambda q}(\lambda q)^{(x-y)}}{(x-y)!}=\lambda q+y \tag{42}$$

which has the form of (2), with $a = 1$ and $b = \lambda q$,

(B) Suppose now that X follows a Binomial distribution with parameters (m,α) and Y|X $= x$ is given by (15). Then E[X|Y $= y$] $= ay + b$, where a and b are constant.

Let be

$$P[X=x,Y=y]=P[X=x]P[Y=y\mid X=x]=\binom{m}{x}\alpha^{x}(1-\alpha)^{m-x}\binom{x}{y}p^{y}q^{x-y}$$
$$=\binom{m}{y}(\alpha p)^{y}\binom{m-y}{m-x}(1-\alpha)^{m-x}(\alpha q)^{x-y} \tag{43}$$

Thus from (43), we have that

$$P[Y=y]=\sum_{x}P[X=x,Y=y]=\sum_{x\geq y}^{m}\binom{m}{y}(\alpha p)^{y}\binom{m-y}{m-x}(1-\alpha)^{m-x}(\alpha q)^{x-y}$$
$$=\binom{m}{y}(\alpha p)^{y}(1-\alpha p)^{m-y} \tag{44}$$

Therefore, Y has a binomial distribution with parameters (m, αp).

Then, from (43) and (44) we have that

$$P[X = x \mid Y = y] = \frac{P[X = x, Y = y]}{P[Y = y]} = \binom{m-y}{m-x}\left(\frac{(1-\alpha}{1-\alpha p}\right)^{m-x}\left(\frac{\alpha q}{1-\alpha p}\right)^{x-y} \quad (45)$$

Therefore, X|Y = y have a binomial distribution with parameters {(m – y), (1 – α)/(1 – αp)}. Then

$$E[X = x \mid Y = y] = \sum_{x \geq y}^{m} x\binom{m-y}{m-x}\left(\frac{(1-\alpha}{1-\alpha p}\right)^{m-x}\left(\frac{\alpha q}{1-\alpha p}\right)^{x-y}$$

$$= \sum_{x \geq y}^{m}(x-y+y)\binom{m-y}{m-x}\left(\frac{(1-\alpha}{1-\alpha p}\right)^{m-x}\left(\frac{\alpha q}{1-\alpha p}\right)^{x-y}$$

$$= (m-y)\left(\frac{\alpha q}{1-\alpha p}\right)\sum_{x \geq y}^{m}\binom{m-y-1}{m-x}\left(\frac{(1-\alpha}{1-\alpha p}\right)^{m-x}\left(\frac{\alpha q}{1-\alpha p}\right)^{x-y-1} + y$$

$$= (m-y)\left(\frac{\alpha q}{1-\alpha p}\right) + y = m\left(\frac{\alpha q}{1-\alpha p}\right) + y\left(1 - \frac{\alpha q}{1-\alpha p}\right) = ay + m(1-a) = ay + b$$

(C) Suppose now that X has a negative binomial distribution with parameters (ν,α), where $\alpha = (1 - ap)/aq$ and Y|X = x is given by (1). Then, E[X|Y = y] = $ay + b$, where a and b are constant.

Let be

$$P[X = x, Y = y] = P[X = x]P[Y = y \mid X = x] = \binom{\nu + x - 1}{x}\alpha^{\nu}(1-\alpha)^{x}\binom{x}{y}p^{y}q^{x-y}$$

$$= \binom{\nu + y - 1}{y}(\alpha)^{\nu}\{p(1-\alpha)\}^{y}\binom{\nu + x - 1}{x - y}\{q(1-\alpha)\}^{x-y} \quad (46)$$

Thus from (46) we have that

$$P[Y = y] = \sum_{x} P[X = x, Y = y] = \sum_{x \geq y}^{\infty}\binom{\nu + y - 1}{y}(\alpha)^{\nu}\{p(1-\alpha)\}^{y}\binom{\nu + x - 1}{x - y}\{q(1-\alpha)\}^{x-y} \quad (47)$$

Therefore, Y has a binomial distribution with parameters (m, αp).

Then, from (46) and (47) we have that

$$P[X = x \mid Y = y] = \frac{P[X = x, Y = y]}{P[Y = y]} = \binom{m-y}{m-x}\left(\frac{(1-\alpha}{1-\alpha p}\right)^{m-x}\left(\frac{\alpha q}{1-\alpha p}\right)^{x-y} \quad (48)$$

Therefore, X|Y = y has a binomial distribution with parameters {(m – y), (1 – α)/(1 – αp)}. Then,

$$E[X = x \mid Y = y] = \sum_{x \geq y}^{m} x \binom{m-y}{m-x} \left(\frac{1-\alpha}{1-\alpha p} \right)^{m-x} \left(\frac{\alpha q}{1-\alpha p} \right)^{x-y}$$

$$= \sum_{x \geq y}^{m} (x - y + y) \binom{m-y}{m-x} \left(\frac{1-\alpha}{1-\alpha p} \right)^{m-x} \left(\frac{\alpha q}{1-\alpha p} \right)^{x-y}$$

$$= (m-y) \left(\frac{\alpha q}{1-\alpha p} \right) \sum_{x \geq y}^{m} \binom{m-y-1}{m-x} \left(\frac{1-\alpha}{1-\alpha p} \right)^{m-x} \left(\frac{\alpha q}{1-\alpha p} \right)^{x-y-1} + y$$

$$= (m-y) \left(\frac{\alpha q}{1-\alpha p} \right) + y = m \left(\frac{\alpha q}{1-\alpha p} \right) + y \left(1 - \frac{\alpha q}{1-\alpha p} \right) = ay + m(1-a) = ay + b$$

which has the form of (16).

8. On characterizing discrete distributions by taking limits in binomial distribution

Let Θ be a continuous r.v. with density function $f(\Theta)$, with $\Theta > 0$ and $E[\theta]$ finite. Let Y be another r.v. whose conditional distribution given $\Theta = \theta$ is given by a Poisson distribution with parameter θ, $P(\theta)$. Then, if $E[\Theta|Y = y] = ay + b$ hold for some constants a,b ($a \neq 0$) if only if Θ follows a Gamma distribution.

Proof ($\Rightarrow$):

Considering the particular case of $E[\Theta|Y = y] = ay + b$ for $y = 0$, we have that:

$$b = E[\Theta = \theta / Y = 0] = \int_0^\infty \theta f(\theta / 0) d\theta > 0 \tag{49}$$

Otherwise, from $E[Y \mid \theta] = \theta$, we have that

$$\theta = E[Y / \theta] = \sum_y y P[Y = y / \theta] = \sum_y y \frac{P[Y = y, \Theta = \theta]}{f(\theta)}$$

from which

$$\theta f(\theta) = \sum_y y P[Y = y, \Theta = \theta] \tag{50}$$

By integrating both sides of (50),

$$\int_0^\infty \theta f(\theta) d\theta = \int_0^\infty \sum_y y P[Y = y, \Theta = \theta] \Rightarrow E[\Theta] = E[Y] \tag{51}$$

Let

$$ay+b=E\left[\Theta/y\right]=\int_0^\infty \theta f(\theta/y)d\theta=\int_0^\infty \theta\frac{P\left[Y=y,\Theta=\theta\right]}{P\left[Y=y\right]}d\theta$$
$$\Rightarrow (ay+b)P\left[Y=y\right]=\int_0^\infty \theta P\left[Y=y,\Theta=\theta\right]d\theta \tag{52}$$

By summing on y both sides of (52) and considering (51), we have that:

$$\sum_y (ay+b)P\left[Y=y\right]=\sum_y \int_0^\infty \theta P\left[Y=y,\Theta=\theta\right]d\theta$$
$$\Rightarrow \; aE\left[Y\right]+b=E\left[\theta\right]\Rightarrow E\left[\theta\right]=b/(1-a)\geq 0$$

Therefore, $a \leq 1$.

From

$$P\left[Y=y\right]=\int_0^\infty \frac{\theta^y}{y!}e^{-\theta}f(\theta)d\theta \tag{53}$$

one obtains that

$$ay+b=E\left[\theta/Y=y\right]=\frac{1}{P(Y=y)}\int_0^\infty \theta\frac{\theta^y}{y!}e^{-\theta}f(\theta)d\theta$$
$$=\frac{y+1}{P(Y=y)}\int_0^\infty \frac{\theta^{y+1}}{(y+1)!}e^{-\theta}f(\theta)d\theta=\frac{(y+1)}{p(Y=y)}P(Y=y+1) \tag{54}$$

From (54):

$$(ay+b)P(Y=y)=(y+1)P(Y+1) \tag{55}$$

Summing both sides of equation (51) and multiplying by t^y, one obtain that

$$atH'(t)+bH(t)=H'(t) \tag{56}$$

whose solution is given by

$$H(t)=\left(\frac{1-a}{1-at}\right)^{b/a}=\left(\frac{p}{1-qt}\right)^{\nu}, \; with\; \nu=b/a \; and\; p=1-a \tag{57}$$

Hence, if $0 < a < 1$, H(t) is the moments generating function of a distribution of probability when ν is a positive real number. In case of a positive integer ν, it is the Negative Binomial distribution, that is, Y ~ $NeBi(\nu,p)$ and when $\nu = 1$, it he Geometric distribution.

Besides

$$H(t)=\sum_y t^y P\left[Y=y\right]=\sum_y t^y\left[\int_0^\infty e^{-\theta}\frac{\theta^y}{y!}f(\theta)d\theta\right]$$
$$=\left[\int_0^\infty e^{-\theta}(\sum_y \frac{(t\theta)^y}{y!})f(\theta)d\theta\right]=\left[\int_0^\infty e^{-(1-t)\theta}f(\theta)d\theta\right]=G(t^*)$$

where $G(t^*)$ is the m.g.f. of θ with $t^* = t - 1$.

By replacing H(t) by $G(t^*)$ in the differentual equation (56):

$$a(t^* + 1)G'(t^*) + bG(t^*) = G'(t^*) \tag{58}$$

whose solution is given by:

$$G(t^*) = \left(\frac{p}{p - qt^*}\right)^{b/q} = \left(\frac{1}{1 - \beta t^*}\right)^{\nu}, \ with\ \nu = b/a = b/q\ and\ p = 1 - a \tag{59}$$

If $0 < a < 1$, then $G(t^*)$ is the generating distribution of a Gamma distribution $\gamma(\nu,\beta)$. Besides, if $\nu = 1$, it is an Exponential distribution and, finally, if $\nu = n/2$ and $\beta = 2$; it is a χ^2 distribution with *n* degrees of freedom.

⇐) To prove the other side, and considering that $Y|\theta \sim P(\theta)$ and $\Theta \sim Ga(\nu,\beta)$, the distribution of (Y,Θ) is given by:

$$P[Y = y, \Theta = \theta] = P[Y/\theta] f(\theta) = \frac{\theta^y}{y!} e^{-\theta} \frac{1}{\Gamma(\nu)\beta^\nu} \theta^{\nu-1} e^{-\theta/\beta} \tag{60}$$

Dividing (60) by the marginal density function of $Y \sim BN(\nu,p)$:

$$f(\theta/y) = \frac{P[Y = y, \Theta = \theta]}{P[Y = y]} = \frac{\frac{\theta^y}{y!} e^{-\theta} \frac{1}{\Gamma(\nu)\beta^\nu} \theta^{\nu-1} e^{-\theta/\beta}}{\binom{\nu + y - 1}{y} p^\nu q^y} \tag{61}$$

$$= \frac{\theta^{\nu+y-1} e^{-\theta/q}}{(\nu + y - 1)!\, q^{\nu+y}} \Rightarrow f(\theta/y) \sim Ga(\nu + y, q)$$

From which it is obtained that $E[\theta \mid y] = \nu + y$, that it is linear, as we wanted to prove.

References

1. Dacey, M.F. (1972). A family of discrete probability distributions defined by the generalized hypergeometric series. Sankhya, Series B, 34, 243–250.
2. Elderton, William P. and Johnson, Norman L. (1969). System of Frequency Curves. Cambridge University Press.
3. Fajardo Caldera, M.A. (1985). Generalizaciones de los sistemas Pearsonianos discretos. Tesis doctoral. Universidad de Extremadura.
4. Hermoso Gutiérrez, J.A. (1986). Estudio sobre distribuciones generadas por funciones hipergeométricas de argumento matricial. Tesis doctoral. Universidad de Granada
5. Herrerías Pleguezuelo, R. (1975). Sobre las estructuras estadísticas de Pearson y exponenciales: problemas asociados. Tesis doctoral. Universidad de Granada.

6. Korwar R.M. (1975). On characterizing some discrete distributions by linear regression. Communications in Statistics, 4(12), 1133–1147.
7. Ord, J.K. (1972). Families of Frequency Distributions. New York: Hafner Publishing Company.
8. Pearson, K. (1920a). "Systematic Fitting of Curves to Observations". New York: Biometrika, 1, 265–303.
9. Pearson, K. (1920b). "Systematic Fitting of Curves to Observations". New York: Biometrika, 2, 1–23.
10. Rodríguez Avi, J. (1993). Contribución a los métodos de generación de distribuciones multivariantes discretas. Tesis doctoral. Universidad de Granada.
11. Sáez Castillo, A.J. (2002). Generación de distribuciones multivariantes discretas mediante funciones hipergeométricas de argumento matricial. Tesis doctoral. Universidad de Granada.

Chapter 6

SOME STOCHASTIC PROPERTIES IN SAMPLING FROM THE NORMAL DISTRIBUTION

J.M. FERNÁNDEZ-PONCE

Departamento de Estadística e I.O., Universidad de Sevilla

T. GÓMEZ- GÓMEZ

Departamento de Estadística e I.O., Universidad de Sevilla

J.L. PINO-MEJÍAS

Departamento de Estadística e I.O., Universidad de Sevilla

R. RODRÍGUEZ-GRIÑOLO

Departamento de Estadística e I.O., Universidad de Sevilla

Univariate stochastic and dispersive ordering have extensively been characterized by many authors over the last two decades. Stochastic orderings are also applied in Economic. In particular, it is interesting to compare situations where one utility function (or one distribution function) is obtained from the other by means of some operation that has an economic meaning. To this end, stochastic properties for distributions associated to normal distribution in sampling are studied in this paper. An application of the multivariate dispersion order in the problem of detection and characterization of influential observations in regression analysis is also shown. This problem can often be associated to compare two multivariate *t*-distributions.

1. Introduction

Stochastic orderings arise in statistical decision theory in the comparison of experiments and estimation problems. Many useful characterizations of the usual stochastic and dispersion order can be found in the literature. An excellent handbook is Shaked and Shantikumar [13]. One of the most interesting characterizations of the dispersion order is given in Shaked [12]. In particular, dispersion and spread has been used to characterize the variability for distributions and it has extensively been studied (see Lewis and Thompson, [10]; Shaked, [12]; Hickey, [8]; Rojo and He, [11]; Fernández-Ponce et al. [6];

among others). An extension of the univariate dispersion order to the multivariate case was given by Giovagnoli and Wynn [7].

Stochastic orderings are also applied in Economic. The typical problem that can be considered is how two different people with two different utilities react to the same uncertain situation and how one person reacts to two different uncertain situations. Stochastic orderings come into play only in the second problem, but the two questions are so deeply related (one is in some sense the dual of the other). In particular, it is interesting to compare situations where one utility function (or one distribution function) is obtained from the other by means of some operation that has an economic meaning.

The paper is organized as follows. In Section 2, the usual stochastic and dispersion order are introduced. It is also given some interesting characterization theorems which will be used later. In Section 3, stochastic properties for distributions in sampling from normal distribution are studied. In Section 4, an application of the multivariate dispersion order in Bayesian Influence Analysis is explained.

2. Univariate stochastic orderings

In this section, the usual stochastic and dispersion ordering are introduced. Moreover, it is given some interesting characterizations theorems which will be used to compare the distributions associated to the normal distribution in sampling.

Definition 2.1. The random variable X is said to be smaller than the random variable Y with respect to the usual stochastic order, denoted as $X \prec_{st} Y$, if $F_X(t) \geq F_Y(t)$ for all $t \in \Re$ or equivalently, if $\overline{F}_X(t) \leq \overline{F}_Y(t)$ for all $t \in \Re$ where $\overline{F}_X$ denotes the survival function of X given by $\overline{F}_X(t) = P(X > t)$.

At first sight it might seem to be counterintuitive to say that $X \prec_{st} Y$ if $F_X(t) \geq F_Y(t)$ for all $t \in \Re$. On the other hand, it is clear that we want to define Y stochastically larger than X when Y has large values with higher probability than X. However, the distribution function describes the probability of assuming small values, hence the reversal of the inequality sign holds. A closure property of stochastic ordering is given in the next theorem.

Theorem 2.1. Let $\{X_i, i = 1, 2, \ldots\}$ be a sequence of non-negative independent random variables, and let M be a non-negative integer valued random variable which is independent of the $X_i's$. Let $\{Y_i, i = 1, 2, \ldots\}$ be another sequence of non-negative independent random variables, and let N be a non-negative integer valued random variable which is independent of the $Y_i's$.

If $X_i <_{st} Y_i$ and if $M <_{st} N$ then $\sum_{j=1}^{M} X_j \prec_{st} \sum_{j=1}^{N} Y_j$.

Proof. See Shaked and Shanthikumar [13].

It seems intuitive that the usual stochastic order can be characterized by using the corresponding density functions. A sufficient condition to order two random variables in the usual stochastic sense is given in the following theorem. A definition is previously needed. Let $a(x)$ be defined on I where I is a subset of the real line. The number of sign changes of a in I is defined by $S^-(a) = \sup S^-[a(x_1), \ldots, a(x_m)]$ where $S^-(y_1, y_2, \ldots, y_m)$ is the number of sign changes of the indicated sequence and the supremum is extended over all sets $x_1 < x_2 < \ldots < x_m$ such that x_i is in I and $m < 1$.

Theorem 2.2. Let X and Y two random variables with density function f and g, respectively. If $S^-(g - f) = 1$ and the sign sequence is −,+ then $X <_{st} Y$.

Definition 2.2. Let X and Y be two random variables with distribution functions F and G, respectively. Let F^{-1} and G^{-1} the right continuous inverses of F and G, respectively. Then X is said to be smaller than Y in the dispersive sense ($X <_{disp} Y$), if $F^{-1}(v) - F^{-1}(u) \leq G^{-1}(v) - G^{-1}(u)$, whenever $0 < u < v < 1$.

In the following theorems, necessary and sufficient conditions are given to order two distributions in dispersion sense. Let supp(X) be the support of random variable X. That is, Supp(X) = $\{x \text{ in } \mathfrak{R}: x > F^{-1}(0)\}$.

Theorem 2.3. $X <_{disp} Y$ if and only if $Y =_{st} \varphi(X)$ for same increasing function φ with $\varphi(x_2) - \varphi(x_1) \geq x_2 - x_1$ for all $x_1 < x_2$ and x_1, x_2 in supp(X). Furthermore, if this is the case then

$$\varphi(x) = G^{-1}F(x) \text{ for all } x \text{ in supp}(X).$$

Proof. See Shaked [12].

Theorem 2.4. The random variable X satisfies $X <_{disp} X + Y$ for any Y independent of X if and only if X has a log-concave density.

Proof. Droste and Wefelmeyer [4].

3. Stochastic properties in normal sampling

In this section, the usual stochastic and dispersion ordering are studied for the distributions associated to the normal distribution in sampling.

3.1. *The normal distribution*

Let X and Y be two random variables with normal distributions $N(\mu_1, \sigma_1)$ and $N(\mu_2, \sigma_2)$, respectively.

Theorem 3.1. Assume that $\sigma_1 = \sigma_2$. $X <_{st} Y$ if and only if $\mu_1 \leq \mu_2$.

Proof.

$$F(t) \geq F(t - \mu_2 + \mu_1) = G(t) \text{ for all } t \text{ in } \Re.$$

Now assume that $\mu_1 = \mu_2$. Then, X and Y can not be compared in usual stochastic sense. See the following example: let X and Y be two normal random variable with density functions $N(0, 1)$ and $N(0, 3)$, respectively. Hence, it is obtained that

$$F(0.5) = 0.69 > G(0.5) = 0.56$$
$$F(1) = 0.15 < G(1) = 0.36$$

Consequently, the usual stochastic ordering is not held.

Theorem 3.2. $X <_{disp} Y$ if and only if $\sigma_1 \leq \sigma_2$.

Proof. By taking into account that the function $\varphi(x) = \frac{\sigma_2}{\sigma_1}(x - \mu_1) + \mu_2$ is an expansion function, the result is immediately obtained by applying Theorem 2.3.

3.2. *The χ^2-distribution*

Let X and Y be two random variables with χ^2-distributions with m and n degrees of freedom, respectively. This fact is denoted as $X \sim \chi^2_m$ and $Y \sim \chi^2_n$.

Theorem 3.3. If $m \leq n$ then $X <_{st} Y$.

Proof. Assume that M and N are random variables with one point distributions on m and n, respectively. Obviously, $M <_{st} N$. Furthermore, assume that X_j and Y_j are independent random variables with the standard normal distribution, then for all j $X_j^2 =_{st} Y_j^2$

Thus, by using Theorem 2.1 is obtained that

$$\sum_{j=1}^{m} X_j^2 \prec_{st} \sum_{j=1}^{n} Y_j^2$$

Theorem 3.4. If $m \leq n$ then $X <_{disp} Y$.

Proof. It is well-known that $X = \sum_{i=1}^{m} Z_i^2$ and $Y = \sum_{i=1}^{n} Z_i^2 = X + \sum_{i=m+1}^{n} Z_i^2$ where $Z_i \sim N(0,\ 1)$. By using that the χ^2-distribution has a log-concave density, the result is obtained by Theorem 2.4.

3.3. *The t-Student distribution*

Let X be a random variable with univariate t-Student distribution and with m and σ degrees of freedom and precision parameter, respectively and denoted as $X \sim St(0, \sigma, m)$. The density function is given by

$$f(x \mid \sigma, m) = \frac{\Gamma\left(\frac{m+1}{2}\right)}{\Gamma\left(\frac{m}{2}\right)(m\pi)^{1/2}} \sigma^{1/2} \left[1 + \frac{\sigma}{m} x^2\right]^{-\frac{m+1}{2}} \text{ for all } x \text{ in } \Re.$$

The standard t-Student distribution, i.e. for σ equals to 1, is denoted by t_m. Now, let X and Y be two random variables with univariate standard t-Student distribution and with m and n degrees of freedom, respectively. It is easy to check that X and Y can not be compared in usual stochastic sense. See the following example: if $X \sim t_2$ and $Y \sim t_5$ then

$$F(-2) = 0.091 > G(-2) = 0.051$$
$$F(1) = 0.788 < G(1) = 0.818.$$

The definition of a univariate partial order strongly connected with the univariate dispersive ordering to order t-Student distributions in dispersion sense is now needed. Therefore, the tail ordering defined by Lawrence [9] is considered. Let X and Y be two univariate random variables symmetric about 0, then we say that X is less in the tail order sense, denoted as $X <_r Y$, if the ratio $G^{-1}(u)/F^{-1}(u)$ is non-decreasing (non-increasing) for u in (1/2, 1) (u in (0, 1/2)). The following theorem establishes a sufficient condition to order the t-Student distributions in dispersion sense.

Theorem 3.3.1. Let $St_1(0, 1, m)$ and $St_1(0, 1, m)$ be two univariate t-distributions. If $n \leq m$ then $t_m <_{disp} t_n$.

Proof. Capéraà [3] showed that if $n \leq m$ then $t_m <_r t_n$. In addition, Doksum [5] showed that for univariate absolutely continuous distributions with $F(0) = G(0) = 0$ such that $f(0) \geq g(0) > 0$ and $G^{-1}(u)/F^{-1}(u)$ non-decreasing for all u in (0, 1), $F <_{disp} G$ holds. Under the last discussion, we consider the random variable $|t_n|$ with density function given by $f_{|t_m|}(t) = 2f_{t_n}(t)$ if $t > 0$ and 0 otherwise. A straightforward computation shows that the distribution function of $|t_n|$ is $F_{|t_n|}(t) = 2F_{t_n}(t) - 1$ for $t \geq 0$. Hence, $F^{-1}_{|t_n|}(u) = F^{-1}_{t_n}\left[(u+1)/2\right]$ for all u in the interval (0,1). Therefore, by using Capéraà [3], if $n \leq m$ then $G^{-1}(u)/F^{-1}(u)$ is non-decreasing for all u in (0, 1). Since $F_{|t_n|}(0) = F_{|t_m|}(0)$ and $f_{|t_m|}(0) > f_{|t_n|}(0)$, by using the result in Doksum [5], $|t_m| <_{disp} |t_n|$ holds. It is easy to check, by using properties of symmetry, that this last result implies that $t_m <_{disp} t_n$.

Note that the degrees of freedom of a t-Student distribution are always associated with the dispersion and the lack of knowledge of the experiment. That is, the lower degrees of freedom the bigger dispersion is and, therefore, the bigger lack of knowledge of the experiment. To study in depth the implications of the univariate dispersion order, see Shaked and Shanthikumar [13]. If the precisions are different, the following corollary holds.

Corollary 3.3.1. Let $St_1(0,\sigma_1,m)$ and $St_1(0,\sigma_2,n)$ be two univariate t-distributions which satisfy that $n \leq m$ and $\sigma_2 \leq \sigma_1$ then $St_1(0,\sigma_1,m) <_{disp} St_1(0,\sigma_2,n)$.

Proof. The proof is trivial and hence is omitted.

3.4. *The F-Snedecor distribution*

Let X be a random variable with an F-Snedecor distribution for n_1 and m degrees of freedom in the numerator and denominator, respectively. This fact is denoted as $X \sim \Im(n_1, m)$. Asume that $X \sim \Im(n_1, m)$ and $Y \sim \Im(n_2, m)$. Then, X and Y can not be compared in usual stochastic and dispersion sense. See the following example: if

$$X \sim \Im(2,10) \text{ and } Y \sim \Im(7,10) \text{ then it is easy to check that}$$

$$F(0.5) = 0.379 > G(0.5) = 0.185$$

$$F(2.5) = 0.868 < G(2.5) = 0.908$$

Note the following plot for the quantiles function is not non-increasing and non-decreasing.

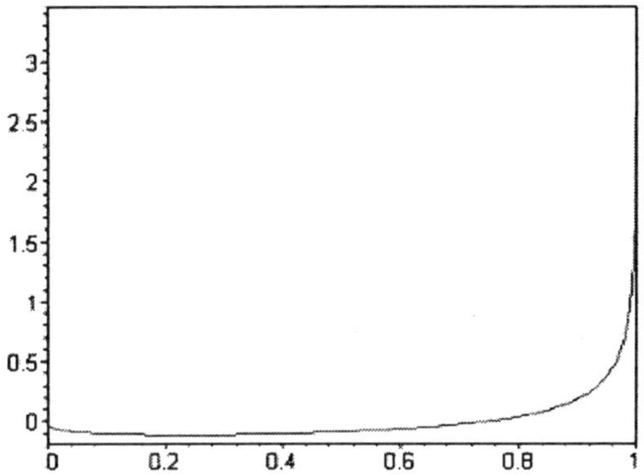

Figure 1. Plot of the function (F^{-1}-G^{-1})

4. An application

In this Section, some results obtained in the last section to the particular *t*-distribution family are applied. For this purpose, the corresponding definition of the *t*-distribution from Bernardo and Smith ([2], pg.139–140) is used. A continuous random vector X has a multivariate *t*-distribution or a multivariate Student distribution of dimension k, with parameters $\mu = (\mu_1,\ldots,\mu_k)$, Σ and n, where μ is in $\Re^k$, Σ is a symmetric positive definite $k \times k$ matrix, and $n > 0$ if its probability density function, denoted $St_k(\mathbf{x};\mu,\Sigma,n)$, is

$$St_k(\mathbf{x};\mu,\Sigma,n) = c\left[1+\frac{1}{n}(\mathbf{x}-\mu)'\Sigma(\mathbf{x}-\mu)\right]^{-\frac{n+k}{2}}$$

for all $\mathbf{x}$ in $\Re^k$ where $c = \dfrac{\Gamma\left(\dfrac{n+k}{2}\right)}{\Gamma\left(\dfrac{n}{2}\right)(n\pi)^{k/2}} |\Sigma|^{1/2}$.

Although not exactly equal to the inverse of the covariance matrix, the parameter Σ is often referred to as the precision matrix of the distribution or, equivalently, the inverse matrix of the dispersion matrix. In the general case, $E[X] = \mu$ and $\mathrm{Var}(X) = \Sigma^{-1}(n/n-2)$. An extension of the univariate dispersion order to the multivariate case was given by Giovagnoli and Wynn [7]. A function $\Phi : \Re^n \to \Re^n$ is called an expansion if $\|\phi(\mathbf{x})-\phi(\mathbf{x}')\|_2 \geq \|\mathbf{x}-\mathbf{x}'\|_2$ for all x and x' in $\Re^n$. Let X and Y be two n-dimensional random vectors. Suppose that $Y =_{st} \Phi(X)$ for some expansion function Φ.Then we say that X is less than Y in the strong multivariate dispersion order (denoted by $X <_{SD} Y$). Roughly speaking, the strong multivariate dispersive order is based on the existence of an expansion function which maps stochastically a random vector to another one. The ordering in the $<_{SD}$ sense is intuitively reasonable and it satisfies many desirable properties. Next result is in order to compare t-distributions when both degrees of freedom and precision matrices are different.

Corollary 4.1. Let $Y_1 \sim St_k(0,\Sigma_1,m)$ and $Y_2 \sim St_k(0,\Sigma_2,n)$ be two multivariate t-distributions with different precision matrices and degrees of freedom. If that $\lambda(\Sigma_2^{-1}) \geq \lambda(\Sigma_1^{-1})$ and $n < m$ hold then $Y_1 <_{SD} Y_2$
where $\lambda(.)$ is the vector of ordered eigenvalues and $\geq$ refers to the usual entrywise ordering.

Proof. See Arias et al. [1].

The following model is considered, $\mathbf{Y} = \mathbf{X}\boldsymbol{\beta} + \boldsymbol{\varepsilon}$, where $\boldsymbol{\varepsilon}$ is an $N\times 1$ random vector distributed as $\mathrm{MN}_n(0,\theta\mathrm{I})$ (N dimensional multivariate normal (MN)) with mean vector zero and covariance matrix $\theta\mathrm{I}$, θ scalar; $\boldsymbol{\beta}$ is the $p \times 1$ vector of regression coefficients; $\boldsymbol{X}$ is an $N \times p$ matrix of fixed "independent" variables; and $\boldsymbol{Y}$ is the $N \times 1$ vector of responses on the "dependent" variable. We assume the prior density for $\boldsymbol{\beta}$ and θ to be $g(\boldsymbol{\beta},\theta) \propto \theta^{-1}$ where $\propto$ means that the first member of this equation is proportional to the second member. This distribution presumes that little prior information is available relative to that information inherent in the data. Assume the case when a particular subset of size k has been deleted, we denote this by **(i)**, while the subset itself is indicated by **i**. Then the general linear model may be expressed as

$$\mathbf{y}' = (\mathbf{y}'_{\mathrm{i}}, \mathbf{y}'_{(\mathrm{i})}) = \boldsymbol{\beta}'(\mathbf{X}'_{\mathrm{i}}, \mathbf{X}'_{(\mathrm{i})}) + (\boldsymbol{\varepsilon}'_{\mathrm{i}}, \boldsymbol{\varepsilon}'_{(\mathrm{i})}).$$

Thus the predictive densities based on full and subset deleted data sets, when θ is unknown, are two multivariate t-distributions with parameters $St_N(\hat{\mathbf{y}}, (\mathbf{s}^2(\mathbf{I}+\mathbf{H}))^{-1}, N-p)$ and $St_N(\hat{\mathbf{y}}^{(\mathrm{i})}, (\mathbf{s}^2_{(\mathrm{i})}(\mathbf{I}+\mathbf{H}^{(\mathrm{i})}))^{-1}, N-k-p)$
where

$$\mathbf{S} = \mathbf{X}'\mathbf{X},\ \ \mathbf{H} = \mathbf{X}\mathbf{S}^{-1}\mathbf{X}',\ \ \mathbf{H}^{(\mathrm{i})} = \mathbf{X}\mathbf{S}_{(\mathrm{i})}^{-1}\mathbf{X}',\ \ \hat{\mathbf{y}} = \mathbf{X}\hat{\boldsymbol{\beta}},$$

$$\mathbf{r} = \mathbf{y} - \hat{\mathbf{y}},\ \ \hat{\mathbf{y}}^{(\mathrm{i})} = \mathbf{X}\hat{\boldsymbol{\beta}}_{(\mathrm{i})},\ \ \mathbf{a}^2 = \mathbf{r}'\mathbf{r},\ \ \mathbf{s}^2 = \mathbf{a}^2 / N - p$$

and $\mathbf{S}_{(\mathrm{i})}, \mathbf{a}^2_{(\mathrm{i})}, \mathbf{s}^2_{(\mathrm{i})}$ are similarly defined.

In this case, the problem to detect influential observations is based on compare two multivariate t-distributions. If we only study the comparison in terms of variability, it seems intuitive that if a subset of data is deleted then the obtained predictive density will be expected to be more dispersive than the predictive density based on full data. That is, the following order is verified

$$f(.) \prec_{SD} f_{(\mathrm{i})}(.)$$

This fact may be interpreted as the added variability, due to deletion of data subset **i**. However it is not held that every subset of data with a fixed size k has the same influence. Consequently, a Dispersion Bayesian Influence in terms of Variability (DBIV) measure to the **i**-subset can be defined as

$$Q_{\mathbf{i}}^2 = \left\| \boldsymbol{\lambda}(\mathbf{s}^2_{(\mathbf{i})}(\mathbf{I}+\mathbf{H}^{(\mathbf{i})})) - \boldsymbol{\lambda}(\mathbf{s}^2(\mathbf{I}+\mathbf{H})) \right\|_2^2$$

and the subsets from least to most influential according to the magnitude of $Q^2{}_{\mathbf{i}}$ are ordered. Note that under the assumptions in Corollary 4.1, if the inequality

$$\boldsymbol{\lambda}(\mathbf{s}^2_{(\mathbf{i})}(\mathbf{I}+\mathbf{H}^{(\mathbf{i})})) \geq \boldsymbol{\lambda}(\mathbf{s}^2(\mathbf{I}+\mathbf{H}))$$

holds then

$$f(.) \prec_{SD} f_{(\mathbf{i})}(.)\,.$$

For more details on this application see Arias et al. [1].

References

1. Arias-Nicolas, J.P. (2005). FernándezPonce, J.M., Luque-Calvo, P. and SuárezLlioréns, A. Multivariate dispersion order and the notion of copula applied to the multivariate t-distribution. Probability in the Engineering and Informational Sciences, 19, 361–375.
2. Bernardo, J.M. and Smith, A.F.M. (1994). Bayesian Theory. John Wiley and Sons.
3. Capéraà, P. (1998). Tail ordering and asymptotic efficiency of rank tests. The Annals of Statistics, 16, 470–478.
4. Droste, W. and Wefelmeyer, W. (1985). A note of strong unimodality and dispersivity. Journal of Applied Probability, 22(1), 235–239.
5. Doksum, K. (1969). Starshaped transformations and the power of rank tests. Annals of Mathematical Statistics, 40, 1167–1176.
6. Fernández-Ponce, J.M., Kochar, S.C. and Muñoz-Perez, J. (1998). Partial orderings of distributions based on right spread functions. Journal of Applied Probability, 35, 221–228.
7. Giovagnoli, A. and Wynn, H.P. (1995). Multivariate dispersion orderings. Statistics and Probability Letters, 22, 325–332.
8. Hickey, R.J. (1986). Concepts of dispersion in distributions: Acomparative note. Journal of Applied Probability, 23, 924–929.
9. Lawrence, M.J. (1975). Inequalities of s-ordered distributions. Ann. Statist., 3, 413–428.
10. Lewis, T. and Thompson, J.W. (1981). Dispersive distributions and the connection between dispersivity and strong unimodality. Journal of Applied Probability, 18, 76–90.
11. Rojo, J. and Guo Zhong He. (1991). New properties and characterizations of the dispersive ordering. Statistics and Probability Letters, 11, 365–372.

12. Shaked, M. (1982). Dispersive ordering of distributions. Journal of Applied Probability, 19, 310–320.
13. Shaked, M. and Shanthikumar, J.G. (1994). Sthocastic Orders and Their Appications. New York: Academic Press.

Chapter 7

GENERATING FUNCTION AND POLARIZATION

R.M. GARCÍA-FERNÁNDEZ

Department of Quantitative Methods in Economics, University of Granada
Campus de Cartuja s/n. Granada, 18071, Spain

In this paper we apply the generating function to obtain the density of the overall sample. This density is called mix density and is proportional to the geometric mean of the subgroups densities. This approach can be use to measure the polarization when it is understood as an economic distance between distributions. An empirical illustration is provided using the data from the Spanish Household Expenditure Survey corresponding to the regions of Andalucía and Cataluña, elaborated by the Instituto Nacional de Estadística (INE) for the year 1999.

1. Introduction

The main objective of this paper is to extend the economic applications of the generating function concept. The generating function was defined by Callejón [1] considering that the right hand side of the system of Pearson, which is given by:

$$\frac{f'(y)}{f(y)} = \frac{y-a}{b_0 + b_1 y + b_2 y^2}$$

is a function of real variable $g(y)$, that is to say $\frac{f'(y)}{f(y)} = g(y)$.

The generating function has been applied successfully to the estimation of the income distribution as we can see for instance in the papers of Herrerías, Palacios and Ramos [8] and Herrerías, Palacios and Callejón [9]. In addition, the concept of generating function can be used to generate Lorenz curves and therefore to measure the inequality of the income distribution [1].

Another economic problem related with the income distribution is the measurement of the polarization of the income as shown by the increasing publications related to this topic (see Esteban and Ray [5], Wolfson [15], Tsui and Wang [13] among others). As we will discuss in Section 4, there are several approaches to measure the polarization. Following Gertel, Giuliodori and Rodriguez [7] we are going to focus on the analysis of the polarization when it is understood as an economic distance between distributions. On this point the properties of the generating function provide a useful frame to the measure of the polarization. Assuming that the income distribution is partitioned in subgroups, by means of the generating function we are going to obtain the density of the overall sample as a geometric normalized mean of the densities of the subgroups. These densities will be used to measure the economic distance between the subgroups distribution.

The approach proposed is developed assuming that the income distribution follows a gamma distribution. We make this assumption because as we can see in empirical studies, the gamma distribution has good properties to fit the income distribution (see among others Lafuente [10] and Prieto [11]). It will be interesting to use other distributions but a full exploration of the different distribution must await a future paper.

This paper is organized as follows. In Section 2 we define the generating function model and obtain the density function of the overall sample as a mix of the density functions of each subgroup. In addition this Section shows how the parameters of the model are estimated. In Section 3, the approach proposed in Section 2 is applied to a gamma distribution. In Section 4, an introduction to the measurement of the polarization is provided focusing on the measure that we are going to use. Section 5, provides an empirical illustration, using the data from the Spanish Household Expenditure Survey corresponding to the regions of Andalucía and Cataluña, elaborated by the Instituto Nacional de Estadística (INE) for the year 1999. The main conclusions are discussed in Section 6.

2. Generating function

The starting point will be the definition of the generating function provided by Callejón [1]. Let Y be a real variable defined over the bounded support (a,b). Suppose that $g(y)$ is a function of real variable such that i) $G(y)=\int g(y)dy$ and ii) $\int_a^b e^{G(y)}dy<\infty$ is verified. Then it is possible to obtain a continuous probability function with density function $f(y)=Ke^{\int^y g(y)dy}$ $(a<y<b)$, in which $K=\left[\int_a^b e^{G(y)}dy\right]^{-1}$. Observe that it is verified:

$$\frac{d}{dy}Lnf(y)=\frac{f'(y)}{f(y)}=g(y) \tag{1}$$

Function $g(y)$ receives the name of generating function of probability (for more details about this function and its properties see Callejón [1]).

Let the support of the distribution be contained in some bounded interval $[a,b]$. Assume that the interval is partitioned into *n* subgroups. Let g_i be the generating function corresponding to each subgroup. It is verified that $g(y)$ can be expressed as the weighted arithmetic mean:

$$g(y) = p_1 g_1(y) + p_2 g_2(y) + ... + p_n g_n(y) \qquad (2)$$

where each weight is a non-negative real number and $\sum_{i=1}^{n} p_i = 1$.

Denoting by f_i the density function associated with the subgroup *i*, and considering expression (1) we can write $f(y)$ as the following normalized geometric mean:

$$f(y) = K f_1(x)^{p_1} f_2(y)^{p_2} ... f_n(y)^{p_n}$$

where *K* is the constant of normalization.

Expressions (1) and (2) allow us to obtain $f(y)$ as a mix of the density functions of each subgroup. This approach, as we will show, can be used to study the degree of polarization presented by the distributions.

3. Application to a gamma distribution

In this Section, we describe the previous process assuming that *Y* follows a gamma distribution of parameters α, ϑ. We make this assumption because our main purpose is to apply this approach to an income distribution, and the gamma distribution, as we can see in empirical studies, has good properties to fit the income distribution (see among others Lafuente [10] and Prieto [11]). Of course, it will be interesting to use other distributions but a full exploration of the different distribution must await a future paper.

To divide the sample, we consider particular characteristics, for example region, occupation... that provide an exhaustive partition of the sample into n subgroups. For simplicity in the exposition, we consider two subgroups whose generating function are $g_1(y;\alpha_1,\vartheta_1)$ and $g_2(y;\alpha_2,\vartheta_2)$ respectively.

The generating functions $g_1(y;\alpha_1,\vartheta_1)$ and $g_2(y;\alpha_2,\vartheta_2)$ of a gamma distribution, are defined in the following form (Callejón [1]):

$$g_1(y;\alpha_1,\vartheta_1) = \frac{\alpha_1 - 1}{y} - \frac{1}{\vartheta_1} \qquad g_2(y;\alpha_2,\vartheta_2) = \frac{\alpha_2 - 1}{y} - \frac{1}{\vartheta_2}$$

According to expression (2) we can write

$$g(y) = p_1\left(\frac{\alpha_1 - 1}{y} - \frac{1}{\vartheta_1}\right) + p_2\left(\frac{\alpha_2 - 1}{y} - \frac{1}{\vartheta_2}\right)$$

$$= \frac{p_1(\alpha_1 - 1) + p_2(\alpha_2 - 1)}{y} - \left(\frac{p_1}{\vartheta_1} + \frac{p_2}{\vartheta_2}\right) = \frac{\alpha - 1}{y} - \frac{1}{\vartheta} = g(y;\alpha,\vartheta)$$

Therefore the density of the overall sample is given by:

$$f(y) = Ke^{\int^{y} g(y)dy} = \frac{1}{\Gamma(\alpha)\vartheta^{\alpha}} y^{\alpha-1} e^{-\frac{y}{\vartheta}}$$

This density is called mix density and is distributed as a gamma distribution where $\Gamma(\alpha)$ is the gamma function.

Observe that the mix density function is proportional to the geometric mean of the densities of the subgroups:

$$f(y) = K\left(\frac{1}{\Gamma(\alpha_1)\vartheta_1^{\alpha_1}} y^{\alpha_1 - 1} e^{-\frac{y}{\vartheta_1}}\right)^{p_1}\left(\frac{1}{\Gamma(\alpha_2)\vartheta_2^{\alpha_2}} y^{\alpha_2 - 1} e^{-\frac{y}{\vartheta_2}}\right)^{p_2} \tag{3}$$

where K is a constant of renormalization given by:

$$K = \frac{1}{\left(\frac{1}{\Gamma(\alpha_1)\vartheta_1^{\alpha_1}}\right)^{p_1}\left(\frac{1}{\Gamma(\alpha_2)\vartheta_2^{\alpha_2}}\right)^{p_2}\Gamma(p_1\alpha_1 + p_2\alpha_2)}\left(\frac{p_1}{\vartheta_1} + \frac{p_2}{\vartheta_2}\right)^{(p_1\alpha_1 + p_2\alpha_2)} \tag{4}$$

Introducing (4) into (3) the mix density function can be rewritten as follows:

$$f(y) = \frac{1}{\Gamma(\alpha)\vartheta^{\alpha}} y^{\alpha-1} e^{-\frac{y}{\vartheta}}$$

where $\alpha = p_1\alpha_1 + p_2\alpha_2$ and $\vartheta = \dfrac{1}{\left(\frac{p_1}{\vartheta_1} + \frac{p_2}{\vartheta_2}\right)}$

In relation to the empirical work, it is necessary to estimate the parameters of the density that is, $\alpha_1, \vartheta_1, \alpha_2, \vartheta_2, p_1, p_2$. We are going to follow the following steps. First, the parameters of the densities $f_1(y;\alpha_1,\vartheta_1)$ and $f_2(y;\alpha_2,\vartheta_2)$ are estimated, using the Method of the Maximum Likelihood Estimation (MLE). That is, we obtain the values of the parameters that maximize the following loglikelihood functions:

$$\ln L(y_1,\ldots,y_n;\alpha_1,\vartheta_1) = -n_1 \ln\Gamma(\alpha_1) - n_1\alpha_1 \ln\vartheta_1 + (\alpha_1 - 1)\sum_{i=1}^{n} \ln y_i - \frac{n_1\bar{Y}_1}{\vartheta_1}$$

$$\ln L(y_1,\ldots y_n;\alpha_2,\vartheta_2) = -n_2 \ln\Gamma(\alpha_2) - n_2\alpha_2 \ln\vartheta_2 + (\alpha_2 - 1)\sum_{i=1}^{n} \ln y_i - \frac{n_2\bar{Y}_2}{\vartheta_2}$$

where n_1 and n_2 are the sizes of the two subgroups and $\bar{Y}_1$ and $\bar{Y}_2$ are the respective sample means.

The values of the parameters that maximize the above loglikelihood function are denoted by $\hat{\alpha}_1, \hat{\vartheta}_1, \hat{\alpha}_2, \hat{\vartheta}_2$. Secondly, we introduce $\hat{\alpha}_1, \hat{\vartheta}_1, \hat{\alpha}_2, \hat{\vartheta}_2$ into $f(y)$, and apply again the Method of the Maximum Likelihood to estimate p_1, p_2. The empirical works show that $\hat{p}_1, \hat{p}_2$ are good for approaching the group size. Observe the parameters α and ϑ can be expressed as a function, $h_i(.)$ of the parameters $\alpha_1, \vartheta_1, \alpha_2, \vartheta_2, p_1, p_2$, that is

$$\alpha = h_1(\alpha_1, \alpha_2, p_1, p_2)$$
$$\vartheta = h_2(\vartheta_1, \vartheta_2, p_1, p_2)$$

Hence, accordingly with the Zenha Theorem (Rohatgi [12]) we can conclude that

$$\hat{\alpha} = h_1(\hat{\alpha}_1, \hat{\alpha}_2, \hat{p}_1, \hat{p}_2)$$
$$\hat{\vartheta} = h_2(\hat{\vartheta}_1, \hat{\vartheta}_2, \hat{p}_1, \hat{p}_2)$$

are the MLE of the parameters α and ϑ.

After describing the estimation process, we are going to apply these results to the polarization measurement.

4. Polarization of the income distribution

First of all, it is necessary to point out, that in this Section, we do not pretend to make an exhaustive study of the polarization measurement. We think that the method proposed in this paper could be useful to analyze the group polarization but this paper is a first approach and it is necessary to continue working on this theme.

Let us first start by defining the notion of polarization. According to Esteban and Ray [5] "in any given distribution of characteristic we mean by polarization the extent to which population is clustered around a small number of distant poles". Several measures of polarization have been defined according to different approaches emphasizing the differences between inequality and polarization. Wolfson [14] proposed the following measure of polarization based on the Lorenz curve:

$$W = 4\frac{\mu}{m}\left[\frac{1}{2} - L\left(\frac{1}{2}\right) - \frac{GI}{2}\right]$$

where μ is the mean, m is the median income, $L\left(\frac{1}{2}\right)$ is the Lorenz curve at the median income and GI is the Gini index.

Tsui and Wang [13] following the measure of Wolfson, defined a new class of indices expressed by:

$$TW = \frac{\vartheta}{N}\sum_{i=1}^{k} n_i \left|\frac{m_i - m}{m}\right|^r$$

where n_i is the number of individuals that belong to group *i*, *k* is the number of groups, m_i is the median of the group *i*, ϑ is a positive constant and r takes values in the interval $[0,1]$.

Esteban and Ray [5] provided a measure of polarization based on the sum of antagonisms between individuals that belong to different groups. The antagonism felt by each individual of group *i* is the joint result of the inter group alienation combined with the sense of identification with the group to which individual *i* belongs. The measure proposed by these authors is:

$$P = \sum_{i=1}^{n}\sum_{j=1}^{n} p_i^{1+\alpha} p_j \left|y_i - y_j\right| \quad 1 \le \alpha \le 1.6$$

where $\left|y_i - y_j\right|$ represents the alienation (distance) felt by the individuals of income y_i and y_j. The share of population given by p_i, and p_i^{α} represents the sense of group identification of each of the p_i members of group *i* within their own group. The parameter α falls into the interval $[1,1.6]$ to be consistent with the set of axioms proposed by Esteban and Ray. Before applying this measure it is necessary to arrange the population into group according to characteristics, for instance, region, race, etc.

Esteban, Gradín and Ray [6] proposed an extension of the Esteban and Ray measure which corrected the error that may appear when the distribution is pre-arranged into groups.

As we can see, the previous measures are defined for the discrete case. A recent paper of Duclos, Esteban and Ray [4] developed the measurement of income polarization which can be described using density function. The measure proposed by these authors is based on what they refer as basis densities, that is, densities unnormalized (by population), symmetric, unimodal and with compact support. It has the following expression:

$$P_\alpha(f) = \int\int f(x)^{1+\alpha} f(y)\left|y - x\right| dydx \quad \alpha \in [0.25, 1]$$

where $\left|y - x\right|$ represents the alienation (distance) felt by the individual located at *x* and *y*. The sense of group identification that an individual with income *x* feels is given by $f(x)^{\alpha}$, where α is the sensitivity to polarization and falls into the interval $[0.25,1]$, in order to be consistent with the set of axioms proposed by Duclos, Esteban and Ray.

Gertel, Giuliodori and Rodriguez [7] measured the polarization of the income distribution using the relative economic affluence measure (*D*) introduced by Dagum [2] (see also Dagum [3]). To define this measure we need to introduce several definitions. Let P be a population with *n* income units y_i. P is partitioned into *k* sub-populations $P_j\,(j=1,...,k)$ of size n_j, with cumulative distribution function $F_j(x)$, and mean income μ_j. The income level of the *i*-th individual that belongs to the *j*-th group is y_{ij}.

Definition 1. The Gini mean difference, Δ_{jh}, is the mathematical expectation of the absolute difference between the income variables *X* and *Y*:

$$\Delta_{jh} = E(|y-x|) = \int_0^\infty \int_0^\infty |y-x| dF_h(x)\, dF_j(y)$$

Definition 2. The gross economic affluence d_{jh} is a weighted average of the income differences $y_{ji} - y_{hr}$ for each y_{ji} of P_j which is higher than y_{hr} of P_h given that P_j is in mean more affluent than P_h $(\mu_j > \mu_h)$:

$$d_{jh} = \int_0^\infty dF_j(y) \int_0^y (y-x)\, dF_h(x) \tag{5}$$

Definition 3. The first order moment of transvariation p_{jh} between the *j*-th and the *h*-th sub-population (such as $\mu_j > \mu_h$) is:

$$p_{jh} = \int_0^\infty dF_h(y) \int_0^y (y-x)\, dF_j(x) \tag{6}$$

Dagum resolved the integrals (5) and (6) obtaining:

$$d_{jh} = E_j(YF_h) + E_h(YF_j) - E_h(Y)$$
$$p_{jh} = E_j(YF_h) + E_h(YF_j) - E_j(Y)$$

where $E_j(YF_h) = \int_0^\infty yF_h(y)\, dF_j(y)$ and $E_j(Y) = \mu_j$.

Considering definitions (1), (2) and (3) the relative economic affluence measure is defined as follows:

$$D = \frac{d_{jh} - p_{jh}}{\Delta_{jh}} = \frac{\mu_j - \mu_h}{\Delta_{jh}}$$

The Gini mean difference can be written as:

$$\Delta_{jh} = 2d_{jh} + \mu_h - \mu_j$$

Hence, the ratio D can be rewritten as:

$$D = \frac{\mu_j - \mu_h}{2d_{jh} - \mu_j + \mu_h} \tag{7}$$

The ratio *D* is a measure of the degree of proximity of the distributions. It takes values on the interval $[0,1]$. It is zero when $\mu_j = \mu_h$, meaning the distributions completely overlap and equal to one when the distributions are totally separate. Therefore, when polarization is interpreted as a distance between distributions, D can be used to measure the polarization. The higher the values of D, the larger the polarization of the income distribution.

In our opinion the last approach is the most appropriate to analyze the polarization in the context that we are working on. That is, we know the densities of the subgroups, $f_1(y)$ and $f_2(y)$, and we want to see how separate or polarized they are.

In the next stage, we are going to obtain D according to the results provided in Section 3. Let us consider two regions, the first group collects the income data from the individuals that belong to region 1, and the second one the individuals that belong to region 2. The mean income of the two regions are given by μ_1 and μ_2, and we assume that $\mu_2 > \mu_1$. The corresponding densities are:

$$f_1(y) = \frac{1}{\Gamma(\alpha_1)\vartheta_1^{\alpha_1}} y^{\alpha_1 - 1} e^{-\frac{y}{\vartheta_1}} \tag{8}$$

$$f_2(y) = \frac{1}{\Gamma(\alpha_2)\vartheta_2^{\alpha_2}} y^{\alpha_2 - 1} e^{-\frac{y}{\vartheta_2}} \tag{9}$$

Given that $\mu_1 = \alpha_1 \vartheta_1$ and $\mu_2 = \alpha_2 \vartheta_2$ we can write expression (7) as follows:

$$D = \frac{\alpha_2 \vartheta_2 - \alpha_1 \vartheta_1}{2d_{21} + \alpha_1 \vartheta_1 - \alpha_2 \vartheta_2}$$

The gross economic affluence, d_{12}, is given by:

$$d_{12} = \int_0^\infty y F_1(y) f_2(y) dy + \int_0^\infty y F_2(y) f_1(y) dy - \mu_1$$

where $f_1(y)$ and $f_2(y)$ are the densities functions (8) and (9) and $F_1(y)$ and $F_2(y)$ are their cumulative densities functions respectively.

As we can see, the ratio D is expressed in terms of the parameters of the gamma distributions and d_1. In Section 3, we described an approach based on the MSE method to estimate the parameters $\alpha_1, \vartheta_1, \alpha_2, \vartheta_2, p_1, p_2$, so the following step will be to apply this theoretical result to an empirical distribution.

5. Empirical application

We want to point out that the main object of this Section is to show how the method proposed works. This is a preliminary version and we do not pretend to do an exhaustive analysis of the income polarization. We are going to use the data from the Spanish Household Expenditure Survey, Encuesta Continua de Presupuestos Familiares, elaborated by the Instituto Nacional de Estadística

(INE) for the year 1999. We are going to focus on the income per capita of two autonomous regions (Comunidades Autómonas), Andalucía and Cataluña. First, we estimate the density function of Andalucia, $f_1(y)$, and of Cataluña, $f_2(y)$. Secondly the mix density associated with the overall sample is estimated, see Figures from 1 to 4.

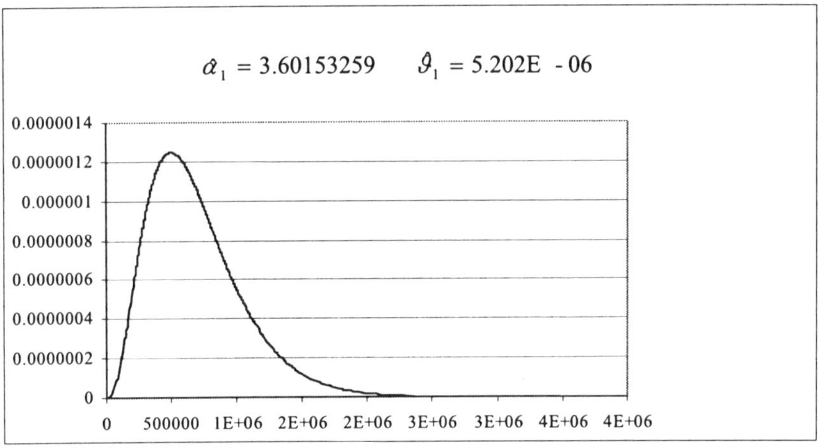

Figure 1. Estimate density function of Andalucía

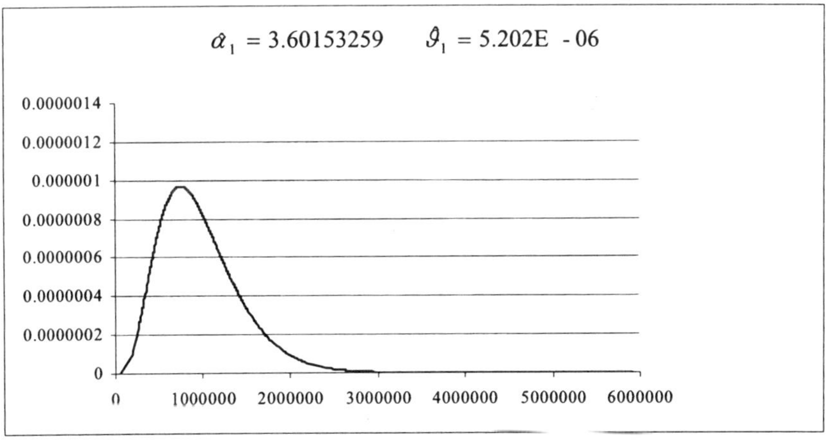

Figure 2. Estimate density function of Cataluña

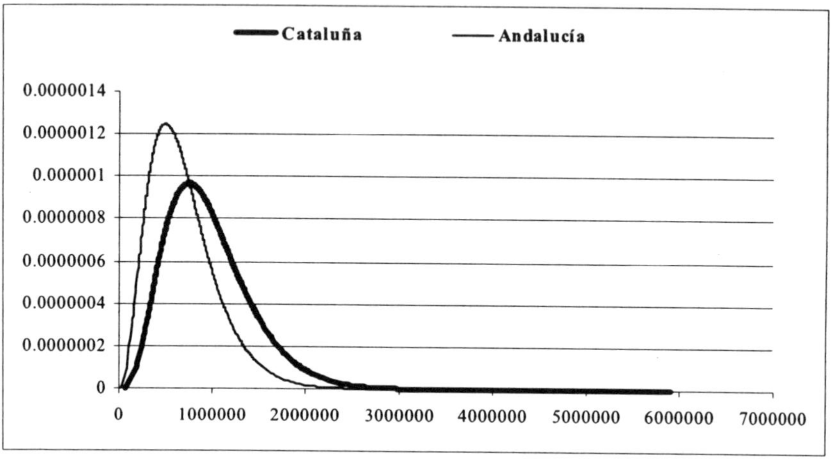

Figure 3. Estimate densities functions of Andalucía and Cataluña

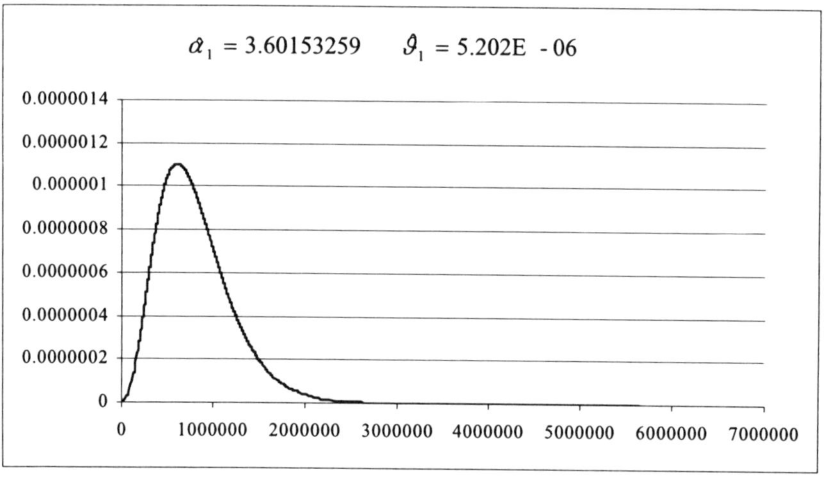

Figure 4. Estimate density function of the mix density

The ratio D can be estimated from the observed values or from a parametric model of income distribution. The estimation presented in this Section is done from the estimated parametric model. To obtain d_{12} we have to resolve by numerical methods the following integrals:

$$\int_0^\infty y\hat{F}_1(y)\hat{f}_2(y)dy + \int_0^\infty y\hat{F}_2(y)\hat{f}_1(y)dy$$

The Gini index of Andalucía and Cataluña, as well as the Gini index for both regions jointly are obtained. Given that the income is distributed according to a gamma distribution, the Gini indices (Lafuente [10]) for Andalucía, IG_1, and Cataluña, IG_2 are calculated using the following expression:

$$IG_i = \frac{\Gamma\left(\alpha_i + \frac{1}{2}\right)}{\sqrt{\pi}\,\Gamma\left(\alpha_i + 1\right)} \qquad i = 1,2$$

The Gini index for the overall sample, considering that $\alpha = p_1\alpha_1 + p_2\alpha_2$, is given by:

$$IG = \frac{\Gamma\left(\alpha_1 p_1 + \alpha_2 p_2 + \frac{1}{2}\right)}{\sqrt{\pi}\,\Gamma\left(\alpha_1 p_1 + \alpha_2 p_2 + 1\right)}$$

The analysis of the Gini index and the ratio *D* jointly, show on the one hand the distance between the income distribution of Andalucía and Cataluña, and on the other hand the inequality in each region. The value taken by *D*, see Table 1, indicates that the income distributions of these two regions are located in an intermediate point between the total overlapping and the complete separation. Concerning the Gini index we conclude that the incomes are more equally distributed in Cataluña than in Andalucía.

As we pointed out at the beginning of this Section, our purpose is to explain how the method development in this preliminary paper works. It will be interesting to obtain the *D* ratio for other years to establish comparison and to consider other characteristics, to group the population, such as education level, occupation etc.

Table 1. Gini indices and *D* ratio

	IG	*D*
Andalucía: $f_1(y)$	0.28718086	
Cataluña : $f_2(y)$	0.25807088	0.554589619
Mix density: $f(y)$	0.27290117	

6. Conclusion and further extensions

First of all, we want to emphasize that the properties of the generating function provide a useful frame to the measure of the polarization when it is understood as an economic distance between distributions. The generating function allows us to obtain the density of the overall sample, which is proportional to the geometric mean of the subgroup densities. This approach makes easy the estimation of the parameters of the mix density. In addition the generating function is a useful tool to extend the measurement of the polarization to antisymmetric densities functions.

The ratio *D* indicates that the income distributions of Andalucía and Cataluña are located in an intermediate point between the total overlapping and the complete separation. In relation to the Gini index we conclude that the incomes are more equally distributed in Cataluña than in Andalucía.

The approach proposed is developed assuming that the income distribution follows a gamma distribution. It will be interesting to use other distributions and to extend the empirical analysis to see how polarization and inequality change over time

References

1. J. Callejón. (1995). Un nuevo método para generar distribuciones de probabilidad. Problemas asociados y aplicaciones. Tesis Doctoral. Universidad de Granada.
2. C. Dagum. (1985). Analyses of income distribution and inequality by education and sex. Advances in Econometrics, 4, 167–227.
3. C. Dagum. (2001). Desigualdad del rédito y bienestar social, descomposición, distancia direccional y distancia métrica entre distribuciones. Estudios de Economía Aplicada, 17, 5–52.
4. J.Y. Duclos, J.M. Esteban and D. Ray. (2004). Polarization: Concepts, measurement, estimation. Econometrica, 74, 1337–1772.
5. J.M. Esteban and D. Ray. (1994). On the measurement of polarization. Econometrica, 62(4), 859–51.
6. J.M. Esteban, C. Gradín C. and D. Ray. (1999). Extensions of a Measure of Polarization OCDE Countries Luxembourg income Study Working Paper 218, New York.
7. R.H. Gertel, R.F. Giuliodori, and A. Rodríguez. (2004). Cambios en la diferenciación de los ingresos de la población del Gran Córdoba entre 1992 y 2000 según el género y el nivel de escolaridad. Revista de Economía y Estadística, XLII.
8. R. Herrerías, F. Palacios and A. Ramos. (1998). Una metodología flexible para la modelización de la distribución de la renta. Décima reunión ASEPELT- ESPAÑA, Actas en CD-ROM.
9. R. Herrerías, F. Palacios and J. Callejón. (2001). Las curvas de Lorenz y el sistema de Pearson 135–151. Aplicaciones Estadísticas y económicas de los sistemas de funciones generadoras. Universidad de Granada.

10. M. Lafuente. (1994). Medidas de cuantificación de la desigualdad de la renta en España según la E.P.F. 1990–91. Tesis Doctoral. Universidad de Murcia.
11. M. Prieto. (1998). Modelización paramétrica de la distribución personal de la renta para España mediante métodos robustos. Tesis Doctoral. Universidad de Valladolid.
12. V.K. Rohatgi. (1976). An Introduction to Probability Theory and Mathematical Statistics. New York: John Wiley and Sons.
13. K. Tsui and Y. Wang. (1998). Polarisation Ordering and New Classes of Polarisation Indices Memo the Chinese University of Hong Kong.
14. M.C. Wolfson. (1994). When inequalities diverge? American Economic Review, 84, 353–58.

Chapter 8

A NEW MEASURE OF DISSIMILARITY BETWEEN DISTRIBUTIONS: APPLICATION TO THE ANALYSIS OF INCOME DISTRIBUTIONS CONVERGENCE IN THE EUROPEAN UNION

F.J. CALLEALTA-BARROSO

Departamento de Estadística, Estructura Económica y O.E.I., University of Alcalá
Plaza de la Victoria no. 2, 28802 Alcalá de Henares (Madrid), Spain

This study introduces a new measure of dissimilarity between distributions, related to Gini's mean difference, and applies it to analyse the convergence between personal income distributions within the 15 EU member states, during the period 1993–2000. According to this measure of dissimilarity, relationships of proximity between these distributions during that period of time constitute the basis of the analysis. Multidimensional scaling techniques are used to construct the temporal trajectories of such distributions in a factor space, optimally reduced for the analysis of their differences. Data are taken from the European Community Household Panel.

1. Introduction

Personal income distribution has been the subject of study from very different perspectives during the last decades. These perspectives have been characterized by terms such as inequality, poverty, deprivation, mobility or convergence.

This study focuses on the measurement of differences between personal income distributions in order to use such a measure as an index of convergence between them.

Measuring these differences raises an important problem for which there is not only one solution. Several interesting aspects can be observed in the personal income distribution of a population, which explains the multiplicity of instruments needed to inform about each of them. Thus, from the simplest descriptive statistics of a distribution to the most sophisticated measures of inequality and poverty, all of them allow us to compare populations in some of their specific aspects.

However, although they achieve successfully the informative specialization for which they were set out, using these measures produces biased results when our aim is to measure the overall difference resulting from the comparison of the individuals that constitute the compared populations. Thus, we can compare the average wealth of two populations from their means, or the internal inequality within them by comparing their Gini's concentration indices. But, for example, in the first case, we are disregarding the information about the shapes of such distributions (it must be remembered that the same mean can be obtained from distributions with different shapes), while in the second case, we are disregarding the localizations of such distributions (it must be remembered that two very different populations can present similar concentration indices, even when one of them can be much richer).

One attempt to avoid this problem is to combine localization statistics with inequality indices. For example, we can consider for this purpose the index $I = \mu \cdot G$, where μ and G are the corresponding mean and the Gini index of the considered distribution, respectively. This index, I, is closely related to Gini's mean difference between the individuals of a population[a].

Could we, therefore, use Gini's mean difference to measure the difference between distributions? Unfortunately, this measure only informs about inter-population inequality[b], and not about proximity[c] between populations. It must be noted that Gini's mean difference between identically distributed populations is not zero but equals twice the product of their common mean and their common Gini index, as can be deduced from footnote [b].

In this paper we propose a new dissimilarity measure related to Gini's mean difference, intuitively interpretable and also clearly informative, which can be used to measure the resulting overall difference between two compared random variable distributions.

---

[a]Let $\Delta = E[|X - Y|]$ be the Gini's mean difference between two random variables X and Y. Then, for X and Y identically distributed, the following equality holds: $I = \mu \cdot G = \Delta/2$.

[b]For any two random variables X and Y, then Δ is related to Gini's inter-population inequality index, G_{xy}, and their localizations, μ_x and μ_y as follows:

$$G_{xy} = \frac{\Delta}{\mu_x + \mu_y}$$

[c]We use the term proximity as a generic reference to any of either dissimilarity or similarity measures, following the terminology used in Cuadras (1996). In order to compare pairs of random variables, (X,Y), this study will concentrate specifically on dissimilarity measures defined as real functions, "*d*", which increase with the difference and comply with the following properties:

$$a) \quad for \quad X = Y \Rightarrow d(X,Y) = 0$$

$$b) \quad d(X,Y) = d(Y,X)$$

These measures are discussed in more detail in Everitt (1993).

Once we have introduced this measure, our objective is to use it to attempt to determine the degree of proximity or convergence that could exist between personal income distributions of different populations over time. Therefore, as an application of what is developed in this paper, we present the study carried out on the convergence between net personal income distributions within the 15 EU member states, during the period 1993–2000, according to the data from the European Community Household Panel (ECHP).

The complexity of the volume of numerical information increases quadratically when we try to address this problem. The dynamic analysis of the degree of convergence between the populations under study requires the measurement of the proximity between them, not only for each period but also throughout the whole period. Thus, to compare p populations over t periods of time we need to take into account

$$\binom{p \cdot t}{2} = \frac{p \cdot t \cdot (p \cdot t - 1)}{2}$$

non-trivial informative indices of proximity calculated between populations for each pair of different periods of time, which have to be interpreted in comparative terms.

This generally large number of informative indices makes it necessary to use a technique beforehand that will allow us to simplify the overall interpretation. We propose, therefore, to apply multidimensional scaling techniques to help us to understand the evolution of distributions in a reduced factor space, whose reference system we will additionally try to explain.

Consequently, for the analysis of the relationships of proximity and distance (convergence) between distributions of net equivalent personal income in the countries under study we will visualize their respective temporal trajectories, which will be found in such a factor space optimally reduced, starting from multidimensional scaling techniques applied to proximity measures previously calculated according to what is proposed in this study.

The problem set out here deals, therefore, with two main issues. On the one hand, we want to find a new measure of dissimilarity, as an informative expression of the degree and quality of the differences observed between the distributions under study. On the other hand, we would like to propose a synthesising methodology for the analysis of these measures, when the objective is to analyse a set of multiple populations through a large number of periods.

2. Measurement of proximity between income distributions

The measure we propose starts from the intuitive idea of the "opulence measure"[d] introduced by Dagum (1980), which he denotes as distance d_1, and which is closely related to Gini's mean difference.

For μ_X and μ_Y the means or average incomes of two populations P_X and P_Y, whose income distributions are represented by random variables X and Y with probability distribution functions $F_X(\cdot)$ and $F_Y(\cdot)$, respectively, Dagum establishes that the population P_Y is more opulent than P_X when $\mu_X < \mu_Y$. In this case, he defines the opulence measure d_1 as follows

$$d_1 = E[(Y-X)\cdot I(Y-X)] = \int_0^{\infty} dF_Y(y) \int_0^{y} (y-x) dF_X(x) \tag{1}$$

where

$$I(Y-X) = \begin{cases} 1 & , \quad Y > X \\ 1/2 & , \quad Y = X \\ 0 & , \quad Y < X \end{cases} \tag{2}$$

Despite the clearly intuitive base of Dagum's proposal, this measure was harshly criticised by Shorrocks (1982), mainly for two reasons:

- Shorrocks considers the measure d_1 inadequate as a relative opulence measure, because Dagum establish, for its calculation, the *a priori* assumption that one of them is more opulent, based exclusively on their mean incomes. Thus, Shorrocks considers that using d_1 as a measurement of the degree of opulence of a population over another might be inconsistent and biased.
- Additionally, Shorrocks considers that d_1 can not be used as a measure of economic "distance" since the measure d_1, applied to compare a distribution to another identically distributed is not zero, as it should logically be. In fact, it equals the product of its mean and its Gini index.

The first observation made by Shorrocks related to Dagum's proposal of prefixing one of the distributions as a reference (that with the bigger mean), once it has been "established" that it is more opulent, also shows a problem when using d_1 as a dissimilarity measure of the difference between distributions.

---

[d]The concept of "opulence" introduced originally by Dagum, corresponds to that of "satisfaction" introduced by Hey and Lambert (1980). The concept of "deprivation" is obtained by changing the role played by both populations. Thus, deprivation of X with respect to Y is defined as opulence of Y with respect to X.

Dagum's proposal introduces a certain economic directionality in his measure, thus making it asymmetrical. Moreover, if we try to use d_1 as a dissimilarity measure, the dissimilarity between the less opulent distribution and the more opulent distribution would not be defined.

However, considering that the intuitive idea that underlies Dagum's measure informs appropriately about the existing economic difference between two distributions, according to Gini's mean difference, we will try to adapt his measure, for our purpose, as we develop below.

2.1. *Reformulation of Dagum's measure d_1 and its relationship to Gini's mean difference*

Gini's mean difference Δ can be re-written as follows:

$$\begin{aligned}\Delta = E\left[\left|Y - X\right|\right] &= E\left[\left|Y - X\right| \cdot I(Y - X) + \left|Y - X\right| \cdot (1 - I(Y - X))\right] \\ &= E\left[\left|Y - X\right| \cdot I(Y - X)\right] + E\left[\left|Y - X\right| \cdot I(X - Y)\right] \\ &= E\left[(Y - X) I(Y - X)\right] + E\left[(X - Y) I(X - Y)\right]\end{aligned} \tag{3}$$

where

$$I(Y - X) = \begin{cases} 1 & , \quad Y > X \\ 0 & , \quad Y \le X \end{cases} \tag{4}$$

According to definitions of opulence and deprivation of a population with respect to another, we could say that two income levels x and y of their respective populations P_X and P_Y support the argument that "P_Y has a greater opulence with respect to P_X" (reciprocally, greater deprivation of P_X with respect to P_Y") if and only if $y>x$. In this case, the amount that this pair of compared levels, (x,y), contributes to the greater opulence of P_Y with respect to P_X, in the sense used by Dagum (reciprocally, to the deprivation of P_X with respect to P_Y), could be evaluated by the difference y-x.

Similarly, we could say that two income levels x and y of their respective populations P_X and P_Y support the argument that "deprivation of P_Y is greater with respect to P_X" (reciprocally, greater opulence of P_X with respect to P_Y") if and only if $y<x$. In this case, the amount that this pair of compared levels, (x,y), contributes to the greater deprivation of P_Y with respect to P_X in the sense used by Dagum (reciprocally, to the greater opulence of P_X with respect to P_Y) could be evaluated by the difference x-y.

The above suggests a decomposition of Gini's mean difference as follows:

$$\begin{aligned}\Delta &= d_{yx}^{+} + d_{yx}^{-} = d_{xy}^{-} + d_{xy}^{+} \\ &= E[(Y-X)I(Y-X)] + E[(X-Y)I(X-Y)]\end{aligned} \tag{5}$$

where:

a) $d_{yx}^{+} = d_{xy}^{-} = E[(Y-X)I(Y-X)]$, is the part of Δ due to mean opulence of P_Y with respect to P_X, which evaluates the difference for the cases in which Y>X. This measure can be interpreted as the mean opulence (satisfaction) of population P_Y with respect to the individuals of P_X with lower incomes (reciprocally, mean deprivation of population P_X with respect to the individuals of P_Y with higher incomes).

b) $d_{yx}^{-} = d_{xy}^{+} = E[(X-Y)I(X-Y)]$, is the part of Δ due to mean deprivation of P_Y with respect to P_X, which evaluates the difference for the cases in which X>Y. This measure can be interpreted as the mean opulence (satisfaction) of population P_X with respect to the individuals of P_Y with lower incomes (reciprocally, mean deprivation of population P_Y with respect to the individuals of P_X with higher incomes).

Given these two definitions, the following properties, which relate them to Gini's mean difference and the means of both compared populations, are satisfied:

- Relationship to Gini's Mean Difference:

$$\Delta = E[|Y-X|] = d_{yx}^{+} + d_{yx}^{-} = d_{xy}^{+} + d_{xy}^{-} \tag{6}$$

- Relationship to the Difference of Means:

$$\mu_Y - \mu_X = E[(Y-X)] = d_{yx}^{+} - d_{yx}^{-} = -(d_{xy}^{+} - d_{xy}^{-}) \tag{7}$$

- Explicit expressions for d^+ and d^-:

$$d_{yx}^{+} = d_{xy}^{-} = \frac{\Delta}{2} + \frac{\mu_Y - \mu_X}{2} \tag{8}$$

$$d_{yx}^{-} = d_{xy}^{+} = \frac{\Delta}{2} - \frac{\mu_Y - \mu_X}{2} \tag{9}$$

- Ranges for d^+ and d^-:

$$0 \le d_{yx}^{+} = d_{xy}^{-} \le \Delta \tag{10}$$

$$0 \leq d_{yx}^{-} = d_{xy}^{+} \leq \Delta \tag{11}$$

Starting from these definitions and properties, Dagum's measure d_1 could be reformulated less ambiguously, as follows:

$$d_1 = \max\left\{d_{yx}^{+}, d_{yx}^{-}\right\} = \frac{d_{yx}^{+} + d_{yx}^{-}}{2} + \frac{\left|d_{yx}^{+} - d_{yx}^{-}\right|}{2} = \frac{\Delta}{2} + \frac{\left|\mu_Y - \mu_X\right|}{2} \tag{12}$$

or, alternatively:

$$d_1 = \max\left\{d_{xy}^{+}, d_{xy}^{-}\right\} = \frac{d_{xy}^{+} + d_{xy}^{-}}{2} + \frac{\left|d_{xy}^{+} - d_{xy}^{-}\right|}{2} = \frac{\Delta}{2} + \frac{\left|\mu_Y - \mu_X\right|}{2} \tag{13}$$

This measure is always between the limits:

$$0 \leq \frac{\Delta}{2} \leq d_1 = \frac{\Delta}{2} + \frac{\left|\mu_Y - \mu_X\right|}{2} \leq \Delta \tag{14}$$

and it takes the maximum value:

$$d_1 = \Delta \quad \Leftrightarrow \quad X > Y \text{ (a.e.) or } X < Y \text{ (a.e.)} \tag{15}$$

We observe that this measure corresponds to the average of two indices of a very different nature, Δ and $\left|\mu_Y - \mu_X\right|$. While $\left|\mu_Y - \mu_X\right|$ summarises the mean difference of wealth, not taking into account the distribution shapes of populations, Δ measures, in absolute terms, inter-population inequality, which appears in the decomposition of the Gini index of two joint populations[e].

With this reformulation, we solve the drawback of asymmetry or unidirectionality presented by the measure of opulence proposed by Dagum when we tried to use d_1 as a dissimilarity measure between both compared populations.

However, the nature of the concentration measure involved in its calculation means that d_1 cannot be considered as a proper measure of dissimilarity. Indeed,

---

[e]When the Gini index is calculated for a population coming from the joining of two others, the part of inequality due to the relationship between the two joint populations after eliminating the part of inequality presented internally by both population separately is:

$$G_{xy} = \frac{\Delta}{\mu_X + \mu_Y}.$$

the measure d_1 of a distribution X to another identically distributed to it is not zero, as it should logically be, but instead:

$$d_1 = \frac{\Delta}{2} = \mu_X \cdot G_X \tag{16}$$

where G_X is the Gini index of X.

With reference to the alternative proposal of relative distance D_1, which Dagum constructs from d_1, consequently, it leads to consider:

$$D_1 = \frac{d_1 - Min(d_1)}{Max(d_1) - Min(d_1)} = \frac{d_1 - \frac{\Delta}{2}}{\Delta - \frac{\Delta}{2}} = \frac{|\mu_Y - \mu_X|}{\Delta} \tag{17}$$

Now, $0 \leq D_1 \leq 1$ and D_1 reaches a minimum value of 0 when means of distributions coincide, not taking into account, in this case, the way in which they distribute their wealth. And it reaches a maximum of 1 as long as any of the variables X or Y are greater than the other (almost everywhere), not taking into account, in this case, their localizations and the distance between their means. This renders it inadequate for our purpose.

However, continuing in the spirit of Dagum concerning this measure and as an attempt to solve the problem presented by using it as a dissimilarity measure, we suggest the following as a new measure of dissimilarity based on the Gini's mean difference between sub-populations of the compared populations.

2.2. *Intuitive approach to a new measure of dissimilarity*

Let X and Y be a pair of absolutely continuous and independent variables with their respective density functions $f_X(x)$ and $f_Y(y)$ defined over $\Re$.

Comparing now the populations, represented by their respective probability density functions, we observe that we can extract two sub-populations, each one coming from each original population, entirely comparable in their values. We can also differentiate them clearly from two other sub-populations, each one coming from each original population, perfectly distinguishable from the other for having "distinctive" values by any individual, as it is intuitively reflected in Figure 1.

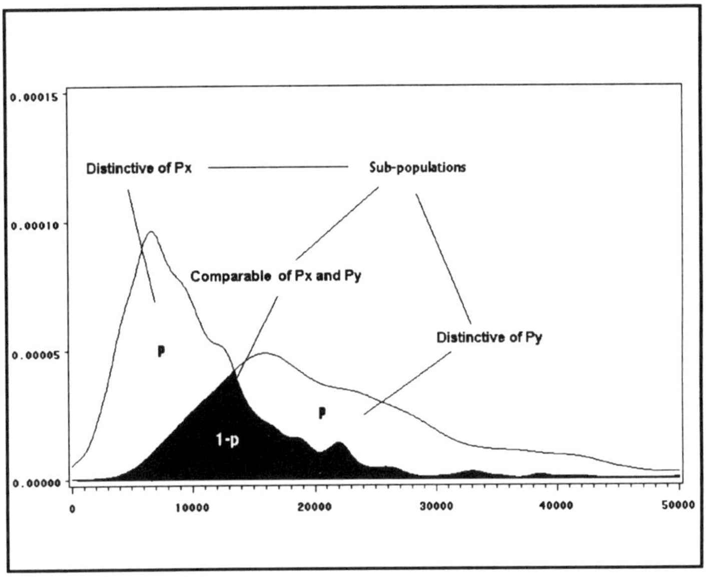

Figure 1. Comparison of populations Px and Py

According to this argumentation, for any pair of absolutely continuous variables, X and Y, the subject of comparison here, we can define the following auxiliary variables:

a) Variable C, which represents the behaviour of the "comparable sub-populations". Here "comparable sub-populations" refer to sub-populations of P_X and P_Y respectively, for each one we can find another sub-population coming from the other population, with similar characteristics; i.e. with similar values of the variable, meaning common behaviour for both variables X and Y (related to the shaded area in Figure 1). Thus, C density function would be set up as follows:

$$f_C(t) = \frac{Min(f_X(t), f_Y(t))}{1-p} = \begin{cases} \dfrac{f_X(t)}{1-p} \quad , & f_X(t) \le f_Y(t) \\ \dfrac{f_Y(t)}{1-p} \quad , & f_Y(t) \le f_X(t) \end{cases} \tag{18}$$

where 1-p is the proportion of each population P_X and P_Y that is "comparable" to another equal proportion in the other one:

$$1 - p = \int_{\{y|f_X(y)>f_Y(y)\}} f_Y(y)dy + \int_{\{x|f_Y(x)>f_X(x)\}} f_X(x)dx \tag{19}$$

b) Variable X^*, which represents the behaviour of the "distinctive sub-population of P_X". Here "distinctive sub-population of P_X" refers to the sub-population of P_X complementary to that selected as "comparable sub-population" to another one of P_Y, with specific characteristics of X and for which it is not possible to find any other element of P_Y "comparable" to its own (related to the non-shaded area on the left hand side, in Figure 1). Its density function would be set up as follows:

$$f_{X^*}(x) = \frac{f_X(x) - f_Y(x)}{p} \cdot I\{f_X(x) > f_Y(x)\}$$
$$= \begin{cases} \dfrac{f_X(x) - f_Y(x)}{p} & , \quad f_X(x) > f_Y(x) \\ 0 & , \quad f_X(x) \le f_Y(x) \end{cases} \tag{20}$$

where p now represents the proportion of population P_X which is "not comparable" to any sub-population of P_Y, and where $I\{\cdot\}$ is the indicative function for the proposition in brackets[f].

c) Variable Y^*, which represents the behaviour of the "distinctive sub-population of P_Y". Here, "distinctive sub-population of P_Y" refers to the sub-population of P_Y complementary to that selected as "comparable sub-population" to another one of P_X, with specific characteristics of Y and for which it is not possible to find any other element of P_X "comparable" to its own (related to the right hand side non-shaded area, in Figure 1). Its density function will be written as follows:

$$f_{Y^*}(y) = \frac{f_Y(y) - f_X(y)}{p} \cdot I\{f_Y(y) > f_X(y)\}$$
$$= \begin{cases} \dfrac{f_Y(y) - f_X(y)}{p} & , \quad f_Y(y) > f_X(y) \\ 0 & , \quad f_Y(y) \le f_X(y) \end{cases} \tag{21}$$

where p now represents the proportion of population P_Y that is "not comparable" to any sub-population of P_X.

With these definitions, the original distributions can be expressed as mixtures of the variables defined above, as follows:

---

[f]The indicative function for a proposition A has a value of

$$I\{A\} = \begin{cases} 1 & , \quad A \text{ true} \\ 0 & , \quad A \text{ false} \end{cases}$$

$$f_X(x) = (1-p)\cdot f_C(x) + p\cdot f_{X^*}(x) \tag{22}$$

$$f_Y(y) = (1-p)\cdot f_C(y) + p\cdot f_{Y^*}(y) \tag{23}$$

where variable C informs about characteristics of the sub-populations selected as "comparable" in both populations P_X and P_Y, with a proportion of 1-p, while the variables X^* and Y^* inform about specific "distinctive" or "non-comparable" sub-populations, of proportions p, coming respectively from either compared populations P_X or P_Y.

Some of these distributions properties are the following:

a) "Distinctive sub-populations" represent a proportion p of populations from which they come, and:

$$p = \frac{1}{2}\cdot\int_{-\infty}^{+\infty}\left|f_X(t) - f_Y(t)\right|\cdot dt \tag{24}$$

b) The means of these auxiliary distributions (C, X^*, Y^*) decompose the means of the original distributions, informing of contributions to the latter of each "comparable" and "distinctive" sub-population, according to their weights in the corresponding mixtures, as follows:

$$\begin{aligned} E[X] &= pE[X^*] + (1-p)E[C] \\ E[Y] &= pE[Y^*] + (1-p)E[C] \end{aligned} \tag{25}$$

From the above we derive the following properties:

$$\begin{aligned} (1-p)E[C] &= E[X] - pE[X^*] = E[Y] - pE[Y^*] \\ E[X] - E[Y] &= p\left(E[X^*] - E[Y^*]\right) \\ E[X] + E[Y] &= p\left(E[X^*] + E[Y^*]\right) + 2(1-p)\cdot E[C] \end{aligned} \tag{26}$$

2.3. *Definition of the proposed measure of dissimilarity*

According to definitions presented above, we propose Gini's mean difference between associated distributions X^* and Y^*, weighted by the product of the proportions they represent of the original populations X and Y[g], as a dissimilarity measure between distributions X and Y.

---

[g]Note that we introduce the weight factor because our objective is, firstly, to make the measure as intuitive as possible (it leads to the direct evaluation of differences related to non-shaded areas in Figure 1). Secondly, we want to introduce in the expression the effect of relative sizes of "distinctive" sub-populations (proportions of populations that "distinctive sub-populations" represent).

$$
\begin{aligned}
d(X,Y) &= p^2 \cdot E\left[\left|Y^* - X^*\right|\right] \\
&= \int_{-\infty}^{+\infty}\int_{-\infty}^{+\infty} |y-x| \cdot (f_X(x) - f_Y(x)) \cdot I\{f_X(x) > f_Y(x)\} \\
&\quad \cdot (f_Y(y) - f_X(y)) \cdot I\{f_Y(y) > f_X(y)\} dx \cdot dy
\end{aligned}
\tag{27}
$$

2.3.1. *Properties of the proposed measure of dissimilarity*

- $d(X,Y) = 0 \Leftrightarrow X = Y \quad (a.e.)$
- The measure *d(X, Y)* increases with the difference between X and Y; i.e., it increases not only with the increase of the proportion of X and Y represented by their "distinctive" sub-populations, but also with the increase of separation between them.
- The measure is symmetrical: $d(X,Y) = p^2 \cdot E\left[\left|X^* - Y^*\right|\right] = d(Y,X)$
- $0 \le d(X,Y) \le \Delta$
- For $X > Y\,(a.e.)$ or $X < Y\ (a.e.) \Rightarrow d(X,Y) = \Delta = \left|\mu_Y - \mu_X\right|$
- This dissimilarity measure is invariant under the same translation of compared variables; and it is proportionally affected by the common scale factor under the same changes of scale for the compared variables.

It is worth noticing, however, that this measure of dissimilarity, which measures proximity between distributions in the way we have proposed, does not strictly fulfil the triangular property; and, therefore, it is not strictly speaking a distance[h].

3. Case study: Convergence of income distributions in the EU-15

3.1. *Concepts and Data*

Following the introduction of the proposed dissimilarity measure, our objective will be to apply it to the analysis of the degree of proximity, and

[h]There are counter-examples in the matrix of dissimilarities calculated in the application developed in a later section. One of them is, for example, that occurring between the countries GER, BEL and FRA in 1993, for which d(GER,FRA)=223 while d(GER,BEL)=59 and d(BEL,FRA)=142.

consequently to the analysis of the convergence, that we may find between personal income distributions within the EU-15 during the last years.

For this purpose, we have used data about family incomes from the European Community Household Panel (ECHP) between 1994 and 2001, which ensures that the information provided is homogenous over time and for the different countries, allowing cross-section and dynamic comparisons.

Looking at sample sizes from the ECHP for each year we can see that we do not have homogenous data for Austria and Luxembourg for the first year (1994), for Finland for the first two years (1994 and 1995), or for Sweden for the first three years (1994–1996). Furthermore, in 1997 Germany, Luxembourg and the United Kingdom (UK) stopped collecting the original ECHP questionnaires, and information requested by ECHP was collected since then from their own national panels (SOEP, PSHELL and BHPS, respectively); we have selected these series of data to preserve longitudinal homogeneity for these countries.

The concept of income used as a starting point is "Total Net Household Income" (variable HI100), which includes incomes after transfer payments and deduction of taxes and Social Security contributions. The years of reference for the ECHP data about incomes correspond to the years before the surveys were carried out and we will use those for this study.

To render incomes comparable for the different countries and waves taken into account, variable HI100 has been adjusted according to purchasing powers of national currencies within each country, using for this purpose the OECD Purchasing Power Parity for each year and currency, also taken from the ECHP (variables PPPyy, yy=93 to 00). And since welfare of a household depends not only on its incomes but also on its size and composition, we have finally calculated the variable "Comparable Equivalent Personal Income (net)", for each country and survey wave, adjusting comparable incomes to this effect.

In short, these adjustments were carried out by dividing "Total Net Household Income", previously modified according to the purchasing power parity for each country and year, by the equivalised size resulting for each household when applying the conventional OECD equivalent scale[i] (variable HD004).

The "Comparable Equivalent Personal Income (net)" has been assigned to each member in every household assuming that all members enjoy the same level of economic welfare. From this approach, the analysis unit is the individual; therefore, in each wave and country the variable "Comparable

[i]In the conventional OECD equivalent scale the first adult counts as 1 unit, next adults as 0.7 and each child under the age of 16 years as 0.5 units.

Equivalent Personal Income (net)" constructed this way, has been weighted by a variable "weight" constructed as the product of the household cross-sectional weight (variable HG004) and its size (variable HD001).

Table 1. Number of available cases for the variable "Comparable Equivalent Personal Income (net)", by countries and waves

	Wave 1 (1994)	Wave 2 (1995)	Wave 3 (1996)	Wave 4 (1997)	Wave 5 (1998)	Wave 6 (1999)	Wave 7 (2000)	Wave 8 (2001)
Germany	6163	6293	6207	6098	5891	5782	5619	5474
Austria	0	3365	3280	3130	2951	2809	2637	2535
Belgium	3454	3341	3189	3009	2857	2684	2549	2322
Denmark	3478	3218	2950	2739	2504	2379	2273	2279
Spain	7142	6448	6132	5714	5438	5299	5047	4948
Finland	0	0	4138	4103	3917	3818	3101	3106
France	7108	6679	6554	6141	5849	5593	5332	5268
Greece	5480	5173	4851	4543	4171	3952	3893	3895
Netherlands	5139	5035	5097	5019	4922	4981	4974	4824
Ireland	4036	3562	3164	2935	2723	2372	1944	1757
Italy	6915	7004	7026	6627	6478	6273	5989	5525
Luxembourg	0	2976	2471	2651	2521	2550	2373	2428
Portugal	4787	4869	4807	4767	4666	4645	4606	4588
U.K.	5024	4987	4991	4958	4958	4914	4842	4749
Sweden	0	0	0	5286	5208	5165	5116	5085

Source: Author's own, from ECHP data

Table 2. Sums of household weights from available cases for the variable "Comparable Equivalent Personal Income (net)", by countries and waves

	Wave 1 (1994)	Wave 2 (1995)	Wave 3 (1996)	Wave 4 (1997)	Wave 5 (1998)	Wave 6 (1999)	Wave 7 (2000)	Wave 8 (2001)
Germany	6140	6280	6207	6125	5921	5812	5646	5506
Austria	-	3366	3280	3133	2954	2809	2636	2539
Belgium	3446	3341	3188	3012	2862	2689	2552	2331
Denmark	3478	3218	2950	2740	2505	2380	2276	2280
Spain	7146	6443	6121	5724	5442	5296	5032	4952
Finland	-	-	4139	4100	3918	3820	3099	3108
France	7113	6683	6564	6141	5853	5596	5333	5277
Greece	5486	5173	4851	4544	4170	3954	3897	3891
Netherlands	5152	5050	5114	5024	4929	4987	4978	4827
Ireland	4038	3565	3164	2938	2725	2374	1947	1759
Italy	6894	6994	7024	6634	6498	6295	6004	5540
Luxembourg	-	2975	2471	2652	2522	2550	2373	2428
Portugal	4799	4868	4809	4780	4653	4655	4614	4592
U.K.	5028	4994	4989	4956	4967	4924	4852	4762
Sweden	-	-	-	5807	5717	5667	5633	5568

Source: Author's own, from ECHP data

Table 3. Weighted means for the variable "Comparable Equivalent Personal Income (net)", by countries and waves (previous year incomes in purchasing parity units)

	Wave 1 (1993)	Wave 2 (1994)	Wave 3 (1995)	Wave 4 (1996)	Wave 5 (1997)	Wave 6 (1998)	Wave 7 (1999)	Wave 8 (2000)
Germany	11479	11424	11959	12429	12727	13102	14012	15166
Austria	-	11917	11887	12070	12159	12579	13686	14359
Belgium	11803	11981	12056	12752	13209	13921	14094	14832
Denmark	11030	11721	12007	12858	13208	13856	14606	14982
Spain	7257	7246	7488	7881	8238	8650	9604	10409
Finland	-	-	9631	10031	10246	10603	10929	11799
France	11022	11052	11224	11358	12116	12673	12709	13549
Greece	6149	6587	6884	7273	7759	7833	8563	8743
Netherlands	10237	10482	11007	11467	12184	12989	13031	13287
Ireland	7702	8849	9481	9614	10860	10672	10709	11616
Italy	8074	8651	8749	8887	9372	9883	10508	10605
Luxembourg	-	18166	18369	19364	19536	20233	21931	23101
Portugal	5898	6270	6377	6719	7018	7432	7792	8619
U.K.	10151	11174	10852	12118	12828	12588	13574	14675
Sweden	-	-	-	10220	10597	10650	11023	12041

Source: Author's own, from ECHP data

Summarizing this first process, Tables 1 and 2 show respectively, by countries and waves, the effective sample sizes and the aggregated sums for the household weights, once we have eliminated the cases for which there is no available data or whose variable "Comparable Equivalent Personal Income (net)" cannot be calculated. Similarly, Table 3 shows the weighted means of variable finally adjusted.

3.2. *Non-parametric estimation of income distributions*

Before calculating the proposed measure of dissimilarity between distributions of "Comparable equivalent personal income" (net), we proceeded to estimate non-parametrically their density functions using univariate gaussian kernels, with optimal bandwidth, following Silverman's procedure (1986) for each country and year considered. For this evaluation we used SAS/STAT procedure KDE[j], which allowed us to calculate the corresponding estimates in each of the 601 equidistant points in which we had divided the common range taken into account (from 0 to 60,000 purchasing parity units), prefixed for all

[j]SAS/STAT® and SAS/GRAPH® are registered products of SAS Institute Inc., Cary, NC, USA.

income distributions in all countries and different waves of the panel[k]. For their analysis, charts for the density functions calculated in this way were obtained using SAS/GRAPH procedure GPLOT.

From these charts we can extract some remarkably different behaviours:

Firstly, we see how Luxembourg has a distribution of "Comparable Equivalent Personal Incomes (net)" clearly displaced to the right of those for the rest of the countries, standing out for its higher personal incomes.

Towards the middle of these charts we can see two other groups of countries behaving differently. The Nordic countries (Finland, Sweden and Denmark) together with Netherlands present distributions more leptokurtic, higher in their central sections (although Denmark and Netherlands present medium degrees of kurtosis). In contrast, the rest of Central European countries present a wider diversity in their central sections of income.

Lastly, on the left hand side of these charts, we find those countries conventionally considered poorer (Italy, Greece, Spain, Portugal and Ireland).

However, if we observe the dynamics of these distributions over time, we can see that although these trends are preserved, most distributions in the EU-15 countries, in general, tend to approach to the others, leaning towards a common average behaviour in the centre of the chart, with the clear exception of Luxembourg and different particularities presented by each country at each period of time.

Additionally, if we observe in these charts the evolution of the distributions for each country through the 8 waves, we can see their systematic movement to the right (a tendency to a higher level of income) with noticeable decreases in modal probability densities (a tendency to a wider diversity of incomes and possibly to a higher inequality) including, in some cases, the presence of central flatness in their density functions and even a pair of relative modes.

We will attempt below to study in depth these first impressions, and for this purpose we will analyse the information obtained from the proposed measure of dissimilarity, calculated between each pair of distributions in all of those.

---

[k]Density functions estimated by stochastic kernels produce small deviations for estimations of population means. Assuming that the ECHP sample sizes have been calculated to obtain parametric estimations rather than for any another reason, we have proceeded to correct slightly the corresponding density function in each case regrouping the upper 1% of probability from the right tail in a unique interval. Thus, we have conveniently determined its range and class-mark so that the mean of the corrected density function would reproduce faithfully the corresponding mean estimated by the ECHP.

3.3. *Dissimilarities*

From the estimates of the 120 density functions obtained in the way mentioned in the previous section (in fact there are 113 since 7 of them are not available, for Austria, Luxembourg, Finland and Sweden, for some of the years) and which represent the behaviours of "Comparable Equivalent Personal Incomes (net)" in the 15 countries studied through the 8 years of the panel, we have proceeded to evaluate the proposed measure of dissimilarity for each pair compared. Consequently, we have constructed the matrix, which reflects the totality of dissimilarity coefficients calculated between every pair of density functions, each one corresponding to a "country-year", using the programme SAS/IML[1].

To sum up, differences between distributions of "Comparable Equivalent Personal Incomes (net)" within the 15 countries for the initial and final years of the period studied are presented in Tables 4 and 5.

Obviously, we cannot calculate the corresponding dissimilarity measures between countries for which data were not available, as it is clearly shown in the table of dissimilarities for the initial year (Table 4). This is the case of Austria, Luxembourg, Finland and Sweden in 1993, Finland and Sweden in 1994 and Sweden in 1995, as mentioned earlier.

Table 4. Dissimilarities between countries for the year 1993

1.993	**GER**	**DK_**	**NL_**	**BEL**	**LUX**	**FRA**	**UK_**	**IRL**	**ITA**	**GRE**	**SPA**	**POR**	**AUS**	**FIN**	**SWE**
GER	0	219	195	59	-	223	147	1.482	953	2.508	1.655	2.885	-	-	-
DK_	219	0	215	202	-	449	484	1.332	1.064	2.447	1.712	2.964	-	-	-
NL_	195	215	0	244	-	117	109	829	526	1.717	911	2.133	-	-	-
BEL	59	202	244	0	-	142	217	1.341	1.024	2.694	1.722	3.006	-	-	-
LUX	-	-	-	-	-	-	-	-	-	-	-	-	-	-	-
FRA	223	449	117	142	-	0	82	947	533	1.842	1.072	2.192	-	-	-
UK_	147	484	109	217	-	82	0	732	301	1.460	730	1.746	-	-	-
IRL	1.482	1.332	829	1.341	-	947	732	0	121	208	43	387	-	-	-
ITA	953	1.064	526	1.024	-	533	301	121	0	436	108	551	-	-	-
GRE	2.508	2.447	1.717	2.694	-	1.842	1.460	208	436	0	144	48	-	-	-
SPA	1.655	1.712	911	1.722	-	1.072	730	43	108	144	0	221	-	-	-
POR	2.885	2.964	2.133	3.006	-	2.192	1.746	387	551	48	221	0	-	-	-
AUS	-	-	-	-	-	-	-	-	-	-	-	-	-	-	-
FIN		-	-	-	-	-	-	-	-	-	-	-	-	-	-
SWE	-	-	-	-	-	-	-	-	-	-	-	-	-	-	-

Source: Author's own, from ECHP data

[1]SAS/IML® is a registered product of SAS Institute Inc., Cary, NC, USA

Table 5. Dissimilarities between countries for the year 2000

2.000	GER	DK_	NL_	BEL	LUX	FRA	UK_	IRL	ITA	GRE	SPA	POR	AUS	FIN	SWE
GER	0	93	243	45	2.665	158	153	765	1.411	3.002	1.583	3.533	90	688	592
DK_	93	0	363	263	2.910	290	374	918	1.432	2.961	1.821	3.652	139	863	780
NL_	243	363	0	96	4.016	35	111	212	439	1.484	674	1.903	130	165	110
BEL	45	263	96	0	3.388	103	112	657	1.078	2.553	1.453	3.062	86	573	488
LUX	2.665	2.910	4.016	3.388	0	4.024	3.039	6.004	7.029	9.403	7.382	9.843	3.222	6.042	5.407
FRA	158	290	35	103	4.024	0	114	250	514	1.587	755	1.993	93	240	203
UK_	153	374	111	112	3.039	114	0	424	884	2.140	1.159	2.612	204	641	424
IRL	765	918	212	657	6.004	250	424	0	61	605	197	1.050	578	117	141
ITA	1.411	1.432	439	1.078	7.029	514	884	61	0	315	77	676	925	171	245
GRE	3.002	2.961	1.484	2.553	9.403	1.587	2.140	605	315	0	171	97	2.306	881	1.016
SPA	1.583	1.821	674	1.453	7.382	755	1.159	197	77	171	0	417	1.264	441	535
POR	3.533	3.652	1.903	3.062	9.843	1.993	2.612	1.050	676	97	417	0	2.750	1.463	1.639
AUS	90	139	130	86	3.222	93	204	578	925	2.306	1.264	2.750	0	505	433
FIN	688	863	165	573	6.042	240	641	117	171	881	441	1.463	505	0	25
SWE	592	780	110	488	5.407	203	424	141	245	1.016	535	1.639	433	25	0

Source: Author's own, from ECHP data

3.4. *Direct analysis of measures of dissimilarity*

In general, we observe a wide range of dissimilarities, going from a few tens of purchasing parity units (43 units for Spain-Ireland in 1993, or 25 units for Finland-Sweden in 2000) to various thousands of units (9,843 units in the case of Luxembourg-Portugal in 2000).

The evolution over time, according to the similarity presented by their distributions, leads to a classification of countries that agrees with that generally found in the economic literature related to the course of these countries.

In Table 6, we have reorganized Table 5 sorting countries in descending order according to their means of the "Comparable Equivalent Personal Income (net)" variable, and set apart the different levels of proximity with different background patterns. Thus, in the year 2000, we would have the following groups of countries with more similar income distributions (some of these countries could be situated alternatively in different contiguous groups, according to the different internal degrees of similarity set up within groups): {Luxembourg}, {Denmark-Germany}, {Germany-Austria-Belgium}, {United Kingdom}, {France-Netherlands}, {Sweden-Finland}, {Ireland-Italy}, {Italy-Spain}, {Greece-Portugal}.

Table 6. Classification of countries in the year 2000, according to their dissimilarities

2000	LUX	DK	GER	AUS	BEL	UK	FRA	NL	SWE	FIN	IRL	ITA	SPA	GRE	POR
LUX	0	2910	2665	3222	3388	3039	4024	4016	5407	6042	6004	7029	7382	9403	9843
DK_	2910	0	93	139	263	374	290	363	780	863	918	1432	1821	2961	3652
GER	2665	93	0	90	45	153	158	243	592	688	765	1411	1583	3002	3533
AUS	3222	139	90	0	86	204	93	130	433	505	578	925	1264	2306	2750
BEL	3388	263	45	86	0	112	103	96	488	573	657	1078	1453	2553	3062
UK_	3039	374	153	204	112	0	114	111	424	641	424	884	1159	2140	2612
FRA	4024	290	158	93	103	114	0	35	203	240	250	514	755	1587	1993
NL_	4016	363	243	130	96	111	35	0	110	165	212	439	674	1484	1903
SWE	5407	780	592	433	488	424	203	110	0	25	141	245	535	1016	1639
FIN	6042	863	688	505	573	641	240	165	25	0	117	171	441	881	1463
IRL	6004	918	765	578	657	424	250	212	141	117	0	61	197	605	1050
ITA	7029	1432	1411	925	1078	884	514	439	245	171	61	0	77	315	676
SPA	7382	1821	1583	1264	1453	1159	755	674	535	441	197	77	0	171	417
GRE	9403	2961	3002	2306	2553	2140	1587	1484	1016	881	605	315	171	0	97
POR	9843	3652	3533	2750	3062	2612	1993	1903	1639	1463	1050	676	417	97	0

Source: Author's own, from ECHP data

Legend:

Dissimilarity less than 100 units
Dissimilarity less than 150 units
Dissimilarity less than 275 units

If we compare the dissimilarities in the final year 2000 to the corresponding dissimilarities in the initial year 1993 we can see which countries have closer distributions at the end of the period than they had at the beginning, and which ones have a greater degree of separation.

To analyse the degree of convergence between countries during this period, we have calculated the convergence indices for each pair of countries, resulting from the ratio of dissimilarity presented by their distributions in the last year of the survey (X_{2000} and Y_{2000}) to that presented in the first year of the survey (X_{1993} and Y_{1993}).

$$I_{1993}^{2000}(X,Y) = \frac{d(X_{2000}, Y_{2000})}{d(X_{1993}, Y_{1993})} \tag{28}$$

Consequently, a value of 1 for this index would show that the distributions compared remain with the same degree of proximity, values greater than 1 would show separation or divergence between the distributions of the countries compared, and values smaller than 1 would show proximity or convergence.

For the cases in which we did not have a dissimilarity measure (in the years 1993, 1994 and 1995) we employed, for the same countries compared, those obtained the following year in which data were available.

Obtained results are shown in Table 7. Starting from it, we can infer that there are groups of countries whose income distributions have come closer during the period 1993-2000. However, there are other countries that present greater differences between them at the end of this period. Consequently, looking at Tables 6 and 7, we can highlight:

a) The country with the highest mean of "Comparable Equivalent Personal Income (net)", Luxembourg, presents a final distribution of incomes clearly distanced from those of the other EU-15 countries.
b) The four countries that follow Luxembourg, according to their mean income (Germany, Denmark, Belgium and the United Kingdom), form a group in which, generally, there is a final greater proximity between income distributions; although with some internal polarizations. Thus, Denmark with Germany and Belgium with the United Kingdom, have respectively reduced their differences to approximately half those presented initially. However, Germany and the United Kingdom practically retain their differences, while Belgium and Denmark have distanced themselves to some extent.
c) Austria has also distanced itself somewhat from the previous countries, with the exception of Denmark, while the latter, has in turn distanced itself from the other two northern countries, Sweden and Finland.
d) Income distributions in these two countries, Sweden and Finland, together with France, Netherlands, Ireland and Italy, become much closer to each other.
e) Out of these countries, Ireland is closer to those with mean incomes higher than its own, with the only exception of Luxembourg which distances itself more rapidly.
f) Spain and Greece (although more so in the case of Spain, which therefore distances itself to a certain extent from Greece) also approach this last group of 6 countries, with the exception of Ireland, which seems to distance itself more rapidly.
g) Income distributions in both Spain and Greece distance themselves from that in Portugal, which seems further from its initial position with respect to the richest countries, shyly approaching the group of 6 countries mentioned in section d, with the exceptions of Italy and, already mentioned, Ireland.

Table 7. Indices of convergence between countries: 1993–2000

Income	Country	LUX	GER	DK	BEL	UK	AUS	FRA	NL	SWE	FIN	IRL	ITA	SPA	GRE	POR
23.101	LUX		1,03	1,19	1,56	1,09	1,43	1,48	1,18	1,14	1,30	1,18	1,37	1,07	1,21	1,17
15.166	GER	1,03		0,42	0,77	1,04	3,13	0,71	1,24	1,47	1,22	0,52	1,48	0,96	1,20	1,22
14.982	DK_	1,19	0,42		1,30	0,77	0,67	0,65	1,69	1,19	1,46	0,69	1,35	1,06	1,21	1,23
14.833	BEL	1,56	0,77	1,30		0,51	2,83	0,73	0,39	1,05	0,92	0,49	1,05	0,84	0,95	1,02
14.676	UK_	1,09	1,04	0,77	0,51		1,40	1,38	1,02	0,99	1,30	0,58	2,94	1,59	1,47	1,50
14.359	AUS	1,43	3,13	0,67	2,83	1,40		1,03	0,54	1,82	0,98	0,55	1,01	0,66	1,00	0,96
13.549	FRA	1,48	0,71	0,65	0,73	1,38	1,03		0,30	0,95	0,67	0,26	0,96	0,70	0,86	0,91
13.287	NL_	1,18	1,24	1,69	0,39	1,02	0,54	0,30		0,81	0,74	0,26	0,83	0,74	0,86	0,89
12.041	SWE	1,14	1,47	1,19	1,05	0,99	1,82	0,95	0,81		0,99	0,26	0,67	0,55	0,85	0,93
11.799	FIN	1,30	1,22	1,46	0,92	1,30	0,98	0,67	0,74	0,99		0,29	0,48	0,51	0,84	0,99
11.616	IRL	1,18	0,52	0,69	0,49	0,58	0,55	0,26	0,26	0,26	0,29		0,50	4,64	2,92	2,71
10.605	ITA	1,37	1,48	1,35	1,05	2,94	1,01	0,96	0,83	0,67	0,48	0,50		0,71	0,72	1,23
10.409	SPA	1,07	0,96	1,06	0,84	1,59	0,66	0,70	0,74	0,55	0,51	4,64	0,71		1,18	1,88
8.743	GRE	1,21	1,20	1,21	0,95	1,47	1,00	0,86	0,86	0,85	0,84	2,92	0,72	1,18		2,03
8.619	POR	1,17	1,22	1,23	1,02	1,50	0,96	0,91	0,89	0,93	0,99	2,71	1,23	1,88	2,03	

Source: Author's own, from ECHP data

Legend:
- Reduction to less than 85% of dissimilarity in 1993
- Reduction to less than 90% of dissimilarity in 1993
- Reduction to less than 95% of dissimilarity in 1993

As can be deduced from the above, the greater or smaller proximity between income distributions in the countries depends not only on the course of the countries' economies, but also on the rhythm or speed with which the other countries move. For this reason, we are not only interested in knowing their current positions but, to a greater extent, to know how they have arrived at it over time: whether trends of proximity (or distance) remained stable throughout that period or not, whether every country tends to a same distribution pattern or not, whether their paths have been relatively similar or not, etc.

To take into account as much information as possible about the course of these income distributions' behaviours, in terms of proximity or distance, we have considered all the dissimilarities calculated between countries' distributions for the years available in the ECHP. Consequently, we have used the totality of the dissimilarity triangular matrix, of the order (8x15)·(8x15), which includes (8x15+1)(8x15)/2=7260 dissimilarity coefficients between the 15 countries' distributions throughout the 8 years of the survey.

As we can see, complexity of numeric information increases since, generally, the measurement of the difference between the behaviours of p populations through t periods, leads us to consider

$$\frac{(p{\cdot}t+1)}{2}\cdot p{\cdot}t = \binom{p{\cdot}t+1}{2}$$

dissimilarity coefficients[m] between p populations for the different t periods of time, which have to be interpreted in comparative terms.

3.5. *Application of the ALSCAL multidimensional scaling model*

To analyse these results and to understand the relative temporal evolution of the distributions studied, it is convenient to treat the dissimilarities previously calculated using a technique that will help us to interpret them globally. For this purpose, we have used a multidimensional scaling technique[n] which will give us a representation of the income distributions compared in an Euclidian factor space of a reduced number of dimensions, deduced optimally according to previously calculated proximity measures. In this space, the analysis of relationships of proximity and convergence between distributions of "Comparable Equivalent Personal Income (net)" in the countries observed will be made easier by the visualization of their respective temporal trajectories. Moreover, once the reference system of this space has been explained in economic terms, the analysis of these trajectories will be more informative.

To simplify the interpretation of dissimilarity coefficients calculated between the distributions of "Comparable Equivalent Personal Income (net)", for each country in each period of time observed, we have used the ALSCAL model, following the procedure established by Young, Lewyckyj and Takane (1986).

This model will attempt to find, in a certain p-dimensional space, the coordinates (points) representative of each country distribution in each survey year, so that the Euclidean distances, d_{ij}, between each pair of these points, or their monotonous transformations $T(d_{ij})$, will reproduce, as closely as possible, the observed dissimilarities, δ_{ij}, between their represented distributions.

[m]Actually, the number of non-trivial dissimilarity coefficients in the matrix resulting from the comparison of the p·t distributions of p countries through t periods, excluding the zero coefficients derived from comparing a country in a period to itself, is:

$$\binom{p{\cdot}t}{2} = \frac{p{\cdot}t{\cdot}(p{\cdot}t-1)}{2}$$

[n]Torgenson (1958) proposed the fundamentals of multidimensional scaling. For an introduction to these methods see Kruskal and Wish (1978).

SAS/STAT procedure MDS[o] has been used in order to solve the adequate ALSCAL model. The model has been established trying a variety of monotonous transformations (identity, afin, lineal, potential and staggered-monotonous), as well as several dimensions for the factor space of representation (between 1 and 6). The goodness of fit criterion used is the measure of Kruskal's Stress-1[p] whose formulation is as follows:

$$S_1 = \left(\frac{\sum_{ij} \left(\delta_{ij} - T(d_{ij})\right)^2}{\sum_{ij} \delta_{ij}^2} \right)^{1/2} \tag{29}$$

and, according to it, we finally find that in all spaces considered, the best approximation was always given by the potential transformation model (or linear logarithmic transformation, equivalently), as follows:

$$\delta_{ij} \cong T(d_{ij}) = s \cdot \left(d_{ij}\right)^p$$

or equivalently,

$$\log\left(\delta_{ij}\right) \cong \log(s) + p \cdot \log\left(d_{ij}\right) \tag{30}$$

For every space of different dimensions considered, this model has provided the values of the goodness of fit criterion reflected in Figure 2, which have also been represented in an elbow chart.

According to these results and following the parsimony principle, two dimensions should be enough to represent quite well the diversity reflected in the calculated dissimilarities; or three if we want the adjustment to be qualified as "excellent", according to Kruskal's scale. Increasing the dimension of representation space to more than three does not seem to improve substantially the goodness of fit for the model, although it improves it to some extent.

Thus, the model has been solved in three dimensions for the potential transformation model (or equivalently, linear logarithmic transformation), obtaining the following optimal solution, whose associated Shepard's Diagram is presented in Figure 3:

$$\delta_{ij} \cong 234.9 \cdot \left(d_{ij}\right)^{1.963}$$

or equivalently,

$$\log\left(\delta_{ij}\right) \cong \log(234.9) + 1.963 \cdot \log\left(d_{ij}\right) \tag{31}$$

[o]SAS/STAT® and SAS/GRAPH® are registered products of SAS Institute Inc., Cary, NC, USA.

[p]SAS/STAT MDS Procedure calculates Kruskal's Stress-1 when options Fit=1, Formula=1 and Coef=Identity are selected. According to Kruskal's criterion, Kruskal's Stress-1 characterizes the goodness of fit of the model as follows: 0=perfect, 0.025=excellent, 0.05=good, 0.1=fair, 0.2=poor. Actually, this is the reason why, in terminology of MDS procedure, it is qualified as a "Badness of Fit Criterion"

Dimensions	Stress-1
1	0.058921
2	0.028915
3	0.022431
4	0.020017
5	0.018404
6	0.017446

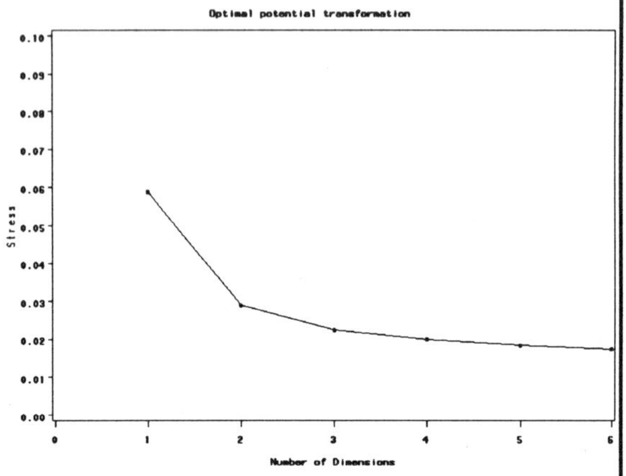

Figure 2. Goodness of fit and dimensionality

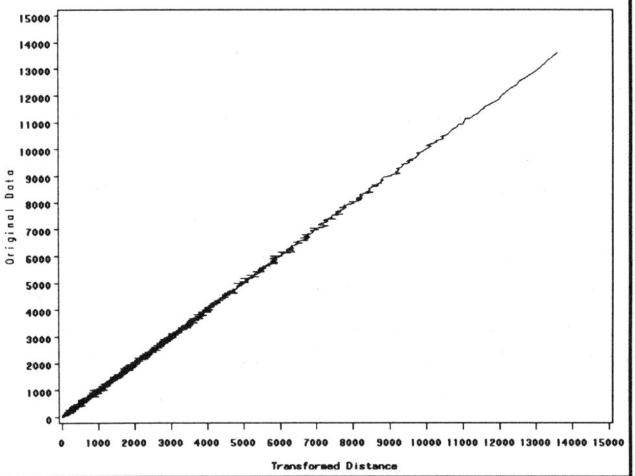

Figure 3. Shepard's diagram

As we can see, Shepard's Diagram[q] confirms the goodness of fit obtained, indicating a very high linear correlation coefficient between the dissimilarities originally observed and the transformations of the corresponding distances

[q]Graphic representation of the pairs $(T(d_{ij}), \delta_{ij})$ joined in order lowest to highest δ_{ij}

reproduced by the coordinates obtained from the model; indeed, once this linear correlation coefficient has been calculated it takes the approximated value of 1.00 with a two decimals precision.

As a consequence, we obtained the coordinates of each country's yearly income distribution in the optimal factor space, which we analyse below.

3.6. *Trajectories of countries' income distributions in the factor space*

Joining in an orderly way the coordinates in the factor space of a specific country throughout the successive years of the survey, we can visualize the trajectory of its behaviour and analyse it comparatively with that of others.

Figure 4 presents the trajectories of Countries' Income Distributions during the period 1993–2000 in the projection plane formed by the two main dimensions of the factor space.

At first glance, we can see that nearly all countries present a quite sustained movement in this period, from right to left along the first dimension, from their position in the initial year to those in the final year, indicated in the chart by the country's identification labels followed by 00.

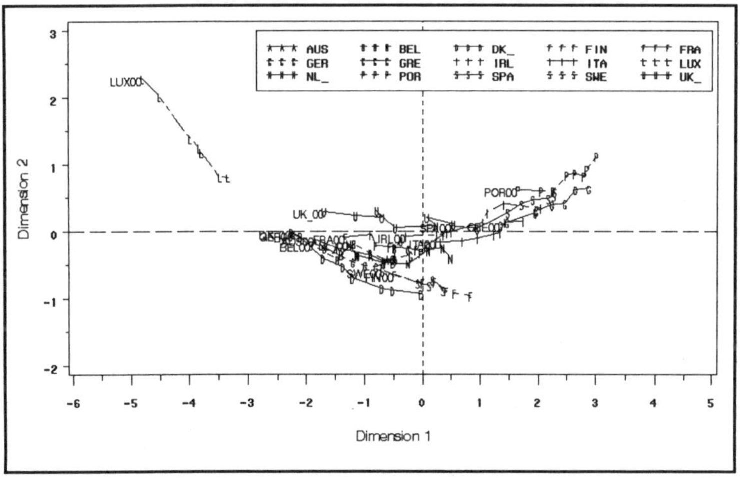

Figure 4. Dynamic of countries using the proposed dissimilarity measure

Source: Author's own, from ECHP 1994–2001

We can also see another generalized movement of concentration of the countries' positions, over time, towards positions close to the reference axis in the first dimension; i.e., towards values close to zero in the second dimension.

There are only two exceptions to this rule: the United Kingdom, whose coordinates seem to raise slightly in the second dimension, although it remains in relatively low levels (+0.29); and Luxembourg, which increases substantially its coordinates in the second dimension, and is far away from the area where trajectories of the rest of the countries are situated.

Although the movement of concentration is generalized with the exceptions mentioned above, we can distinguish five groups of countries with quite different value levels at the end of the period studied: Luxembourg (+2.28), Portugal (+0.68), the United Kingdom (+0.29), Sweden and Finland (–0.59 and –0.65 respectively) and the rest of the countries (between –0.20 and +0.10).

3.7. *Understanding the factor space*

Since our aim is to analyse the totality of the cloud of points in its three dimensions, from their two-dimension projections over each pair of them, we will try to study their statistical and economic interpretation in more depth, in an attempt to better understand what is reflected in the charts.

To this end, Figures 5 and 6 show correlations between punctuations on the dimensions of the factor space and the following descriptive variables (represented in these figures as they appear in brackets): arithmetic average (mean), quantiles of the orders 0.001, 0.01, 0.05, 0.1, 0.25, 0.5, 0.75, 0.90, 0.95, 0.99, 0.999 (P_001, P_01, P_05, P_10, P_25, P_50, P_75, P_90, P_95, P_99, P_999), ratios of these quantiles to the mean (Pr_001, Pr_01, Pr_05, Pr_10, Pr_25, Pr_50, Pr_75, Pr_90, Pr_95, Pr_99, Pr_999), standard deviation (dt), range (rango), interquantile range P_25-P_001 (r1), interquantile range P_50-P_25 (r2), interquantile range P_75-P_50 (r3), interquantile range P_999-P_75 (r4), Pearson's variation coefficient (cvar), ratios of these four interquantile ranges to the median (rr1, rr2, rr3, rr4), Gini's mean difference (dmgini), Gini's concentration index (igini) and squared Pearson's variation coefficient (cvar2).

The graphic representation of these descriptive variables, according to their correlations with the dimensions of the factor space, will allow us to study their intuitive meaning. In order to simplify these graphic representations, we only represent those descriptive variables for which at least one of the correlations with any represented dimension is higher that 0.4. Thus, Figure 5 shows descriptive variables in the first two main dimensions sub-space.

We can see that the first dimension is highly and negatively correlated (correlations near to –1) with nearly all the localization measures, absolute and relative, and also with dispersion measures such as Gini's mean difference and the standard deviation. In addition, it is also positively correlated with Gini's concentration index. Therefore, a country will be located the more to the left of the chart, the more its income distribution moves to the right over the income size axis, providing higher average incomes and distributing higher wealth (which usually happens with an increase of dispersion in the distribution), and the lower inequality it presents (the more evenly the incomes are distributed).

Therefore, the first dimension can be interpreted as an index of welfare or an index of "standards of living-income"[r] which takes into account jointly the general level of wealth in the population and the degree of equality in the way it is distributed.

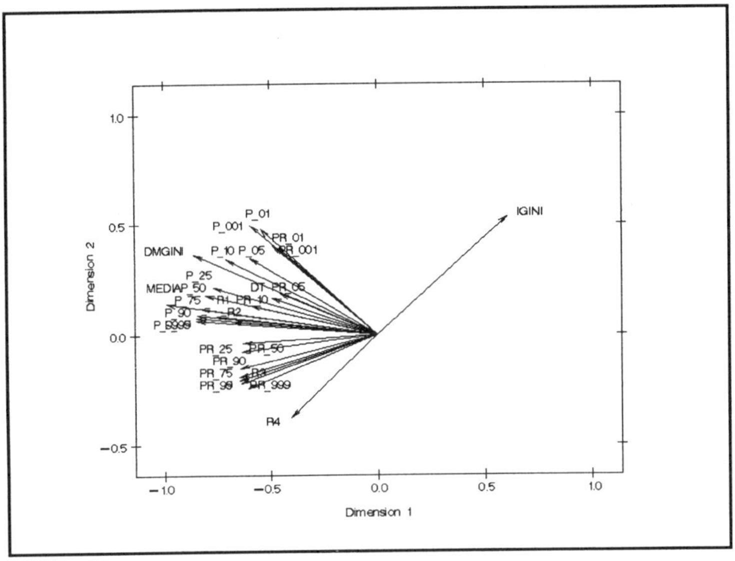

Figure 5. Chart of descriptives in dimensions 1 and 2

Looking now at the second dimension, we observe that its correlations with the descriptive measures are not too high, and therefore its interpretation could be risky. In any case, the descriptive statistic more closely correlated with it is Gini's concentration index (positively correlated), although ratios of low percentiles to the mean are also positively correlated, and ratios of high percentiles to the mean are negatively correlated as well.

[r]The group of "standards of living-income" indices introduced by Pena et al. (1996) is defined as the product of the income distribution mean and the complement to 1 of a normalized inequality index. It belongs to a wider class of welfare indices introduced by Blackorby and Donaldson (1978).

Therefore, this dimension classifies the income distributions of the different countries placing in the more positive values those that have a greater concentration of percentiles around their means and reach at the same time high degrees of inequality. Reciprocally, it places in the more negative values those distributions that have a greater separation of percentiles around their means and reach at the same time low degrees of inequality. Hence, this dimension seems to inform about the contribution of the right distribution tail to inequality. More positive values in this dimension denote countries where the right tail has a higher relative weight in the inequality, to compensate the higher equality in the rest of the distribution, and *vice versa*.

To interpret the third dimension, let us observe the corresponding chart of descriptive variables on the projection plane over dimensions 1 and 3, as shown in Figure 6. We can see that, as was the case with the second dimension, correlations with the third one are not high and, therefore, its interpretation is risky. In any case, the highest correlations in absolute value are negative and correspond to dispersion statistics, especially relative dispersion statistics, and indices of inequality, while all the localization statistics present positive correlations, especially the ratios of low percentiles to the mean. Furthermore, we can see that this dimension is virtually not correlated with the mean.

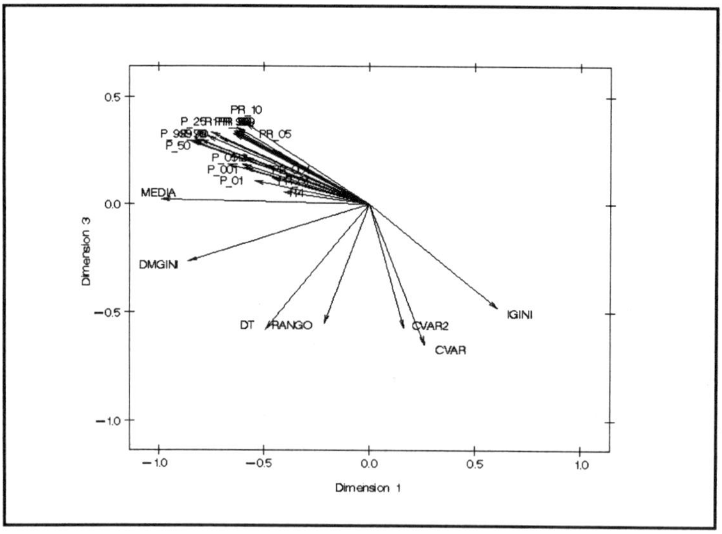

Figure 6. Chart of descriptives in dimensions 1 and 3

Thus, this dimension classifies income distributions in the different countries placing in more positive values those that present a greater distance with respect to the mean of percentiles above this and a smaller distance with respect to the mean of percentiles below this, presenting at the same time a lower inequality and a lower dispersion. Reciprocally, this dimension places in more negative values those distributions that present a greater distance with respect to the mean of percentiles below it and a greater proximity to the mean of percentiles above it, presenting at the same time a greater inequality and a greater dispersion.

Consequently, this dimension seems to inform about the contribution to the inequality of the lower, middle and middle-upper classes or about the structure of incomes in these classes.

The kind of difference that dimension 3 informs about seems to be related to the structures of the left tail and central section, sometimes providing local positive or negative skewness in these sections of the distributions, sometimes favouring flatness or more than one relative mode, and sometimes producing more bell-shaped and symmetric forms.

To sum up, first dimension informs us about welfare in the sense of "standards of living-income", fundamentally influenced by the mean of the population's incomes. The second and third dimensions seem to inform about the different patterns of the same standards of living-income, using different ways of internal distribution of wealth (i.e., different ways of obtaining similar levels of global welfare with different forms of internal inequality).

3.8. *Analysis of trajectories of income distributions*

Bearing in mind the above interpretations of dimensions of the factor space in which we can observe most of the variability between income distributions in the considered countries, during the years observed by the ECHP, we will finally analyse below the trajectories followed by these countries throughout the years. This basic analysis will be carried out on the representations of income distributions in the three projection planes formed by each two of the three dimensions considered, at a larger scale (i.e., without Luxembourg).

Starting from the representations in Figures 7, 8 and 9, we can now complement the analysis from convergence indices carried out in Section 3.4:

a) All countries show a rather sustained movement towards the left on the dimension 1 (positions of greater welfare or standards of living-income). Only the United Kingdom seems to have had a few setbacks during the years 1995 and 1998, compensated largely by its progress during the rest of the period.

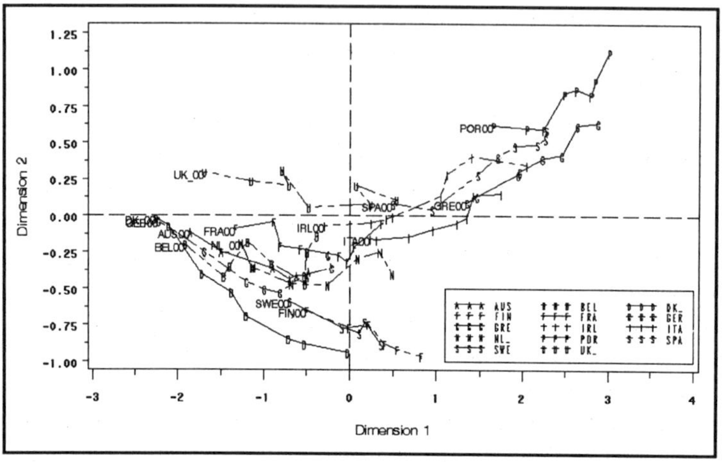

Figure 7. Dynamic of countries (without Luxembourg) using the proposed dissimilarity measure: Dimensions 1 and 2

Source: Author's own, from ECHP 1994–2001

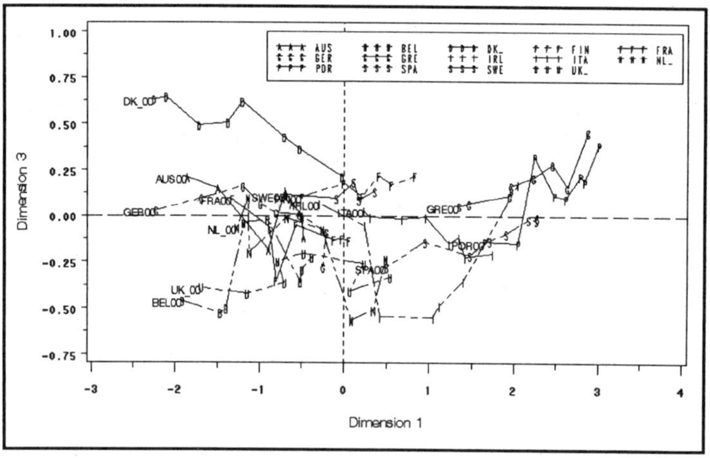

Figure 8. Dynamic of countries (without Luxembourg) using the proposed dissimilarity measure: Dimensions 1 and 3

Source: Author's own, from ECHP 1994–2001

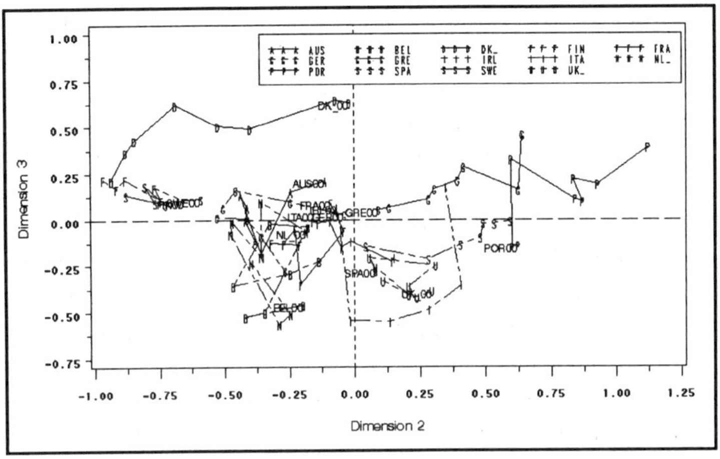

Figure 9. Dynamic of countries (without Luxembourg) using the proposed dissimilarity measure: Dimensions 2 and 3

Source: Author's own, from ECHP 1994–2001

b) Luxembourg, the country with the higher equivalent and comparable mean income, presents a final income distribution clearly distanced from that of the rest of the EU-15. This is due to the inequality caused by the progressively more heavy weight in its right tail (dimension 2) as well as its clearly higher mean of incomes, which compensates its greater inequality and leads it to have the higher "standards of living-income" in the EU-15 (as shown by dimension 1, in Figure 4).

c) The internal polarizations in the convergent group of countries formed by Germany, Denmark, Belgium and the United Kingdom, identified in the analysis of convergence indices, are due to the different ways in which they distribute their wealth. Despite the fact that Denmark with Germany and Belgium with the United Kingdom converge in very similar standards of living-income (dimension 1), they differ in the way in which they distribute their wealth. Particularly, in the first case, in dimension 3 (inequality due to the low and central areas of their distributions) and in the second case, in dimension 2 (inequality due to the right tail of the distribution).

d) Despite the exceptional behaviour of Luxembourg's income distributions, quite different from that in the rest of the EU-15, virtually all countries have a tendency towards average values (close to zero) in the second dimension, with the only exception of the United Kingdom.

Those countries tend to converge, therefore, in a unique model according to the kind of inequality that characterises this dimension (with a right tail of medium weight).

e) However, the United Kingdom increases its value in the second dimension since 1996, slightly but in a sustained way, remaining within values of around +0.30.

f) In any case, according to the inequality that characterizes the second dimension (heaviness on the right tail), we find different states of mutual proximity at the end of the period analysed, as a consequence of this convergence process: Luxembourg (+2.28), Portugal (+0.68), the United Kingdom (+0.29), Sweden and Finland (−0.59 and −0.65 respectively, around a value of −0.60) and the rest of the countries (between −0.20 and +0.10), where higher positive values indicate higher relative heaviness of right tails.

g) Austria has distanced itself to some extent from Germany and Belgium. But this distance is due to their different behaviours in dimension 3 (inequality due to the lower and central sections of their distributions) since they started off from a similar position in the first year, in dimensions 1 and 2, and they have only distanced themselves very slightly in absolute terms.

h) With respect to dimension 1, dimension 3 seems to take a "U" shape, for most of the countries. Dimension 3 decreases in countries with lower levels of welfare as they increase them (increasing the part of inequality due to the enlargement of lower and middle sections of the income distribution). Dimension 3 increases in countries with higher levels of welfare (decreasing the part of the inequality due to the lower and central sections of the income distribution). Exceptions to this rule are Belgium and the United Kingdom, whose values decrease following corrections in this dimension for the year 1997, while Sweden and Finland remain at stable levels.

i) Denmark, Sweden and Finland share trends of growth in the first two dimensions towards a distribution pattern, which could be referred to as "Central-European". Denmark, in particular, shows a greater impetus, with higher growth in welfare and inequality related to the second dimension (weight in its right tail), and therefore distances itself more. However, inequality in the lower and middle sections of income distributions in Sweden and Finland also increases, while global inequality in Denmark is compensated to a certain extend with greater equality in these sections (according to third dimension).

j) Income distributions in France, Netherlands, Ireland and Italy have also come closer, with levels of inequality approximately similar in both dimensions (2 and 3). Sweden and Finland tend to converge to them, although the final dissimilarities in the second dimension continue to be greater than in these four countries.

k) Starting off with high inequality levels in the second dimension, Ireland, Spain and Greece have decreased their inequalities to average European levels. But this is not the case with the inequality associated to the third dimension, which increases in the middle-lower section, even though in Ireland it changes its trend in 1995-96. In any case, Ireland is the country that gets closer out of all countries with higher mean of incomes than itself, not only in welfare but also in levels of inequality. The only exception is Luxembourg, which distances itself more rapidly.

l) Spain and Greece (although more so in the case of Spain) also get closer to the group of Sweden, Finland, France, Netherlands, Ireland and Italy (but Ireland gets closer to the richer countries more rapidly and increases its distance from Greece and Spain). The growth of inequality in the third dimension for Spain implies levels of inequality in its middle and lower classes above the average of the EU-15.

m) Spain and Greece distance their income distributions with respect to that of Portugal, which maintains levels of inequality above the EU-15 average, in both dimensions. This occurs despite the fact that inequality is reduced in the second dimension because of its increase in the third dimension. Regarding welfare, Portugal seems to get even further than initially with respect to the richer countries, getting slowly closer to the group of six countries referred to above (in j) with the exception of Ireland, as mentioned earlier.

4. Conclusions

Taking as a starting point the problematic proposal made by Dagum (1980) to measure the distance between income distributions, we have introduced in this study a new measure of dissimilarity, based on Gini's mean difference.

To test its validity we have calculated the corresponding measures of dissimilarity between all the yearly distributions of "Comparable Equivalent Personal Incomes (net)" in the EU-15. Distributions and dissimilarities have been constructed on the basis of the data from the EU-15 Household Panels between 1994 and 2001.

The analysis of each yearly table of dissimilarities between countries thus calculated, allows to describe the relative situation in the countries for each year. In this way, we have analysed the data for the last year of reference about incomes (2000) and established the groups of more similar countries in that year.

Since these results are a consequence of an evolutionary process over time, comparison of tables of dissimilarities from two different years allows us to analyse the transformation experienced in the period of time studied. We have, therefore, extracted some specific consequences based on the magnitudes of the proximity relationships, between the different countries, by comparing their position in the final year (2000) with that of the initial year (1993). Thus we have determined the groups of countries whose distributions have come closer (converge) and those whose distributions have mutually distanced (diverge).

However, this static comparison ignores the dynamic of the process, i.e. the evolution of the countries to reach the point of the final transformation observed, and does not reflect the possible sources of diversity which could explain the different behaviours of the studied income distributions. In order to visualise these dynamics, we have applied the ALSCAL multidimensional scaling model to determine, first of all, in which dimensions the differences between distributions manifest themselves (dimensions of a factor space). We have concluded, in short, that they are fundamentally three: standard of living-income, inequality due to heaviness on the right tail of the distribution, and inequality due to the lower and middle classes of the distribution. The model has also been applied to describe the trajectories followed by the distributions of the countries considered, in that factor space.

The conclusions drawn from this analysis, applied to the general and specific dynamic of the set of considered countries, have been detailed in the sections 3.4 and 3.8; and we refer the reader to these sections to avoid repetition here.

Acknowledgments

This study has been partially supported by Project I+D+I ref.: SEC2002-00999, from the Spanish Ministerio de Ciencia y Tecnología. Data from the European Community Household Panel have been used here by permission given in the agreement ECHP/15/00, between EUROSTAT and the University of Alcalá (Spain).

References

1. C. Blackorby and D. Donaldson. (1978). Measures of relative equality and their meaning in terms of social welfare. Journal of Economic Theory, 18, 59–80.
2. C.M. Cuadras. (1996). Métodos de Análisis Multivariante. EUB.
3. C. Dagum. (1980). Inequality measures between income distributions with applications. Econometrita, 48(7), 1971–1803.
4. Eurostat. (2004). ECHP UDB Manual: European Community Household Panel Longitudinal Users' Database. Eurostat.
5. B.S. Everitt. (1993). Cluster Analysis. New York: John Wiley and Sons.
6. C. García, F.J. Callealta and J.J. Núñez. (2005). La Interpretación Económica de los Parámetros de los Modelos Probabilísticos para la Distribución Personal de la Renta. Una Propuesta de Caracterización y su Aplicación a los Modelos de Dagum en el Caso Español. Estadística Española, I.N.E.
7. Hey and Lambert. (1980). Relative deprivation and the Gini coefficient: comment. Quarterly Journal of Economics, 95, 567–573.
8. J.B. Kruskal and M. Wish. (1978). Multidimensional Scaling. Sage University Paper series on Qualitative Applications in the Social Sciences, 7–11. Sage Publications.
9. B. Pena, F. J. Callealta, J. M. Casas, A. Merediz and J. J. Nuñez. (1996). Distribución Personal de la Renta en España. Pirámide.
10. B.W. Silverman. (1986). Density Estimation for Statistics and Data Analysis. London: Chapman and Hall.
11. A.F. Shorrocks. (1982). On the distance between income distributions. Econometrica, 50(5), 1337–1339.
12. W.S. Torgenson. (1958). Theory and Methods of Scaling. John Wiley and Sons, Inc.
13. J. Villaverde Castro and A. Maza Fernández. (2003). Desigualdades Regionales y Dependencia Espacial en la Unión Europea. CLM Economía, 2, 109–128.
14. F.W. Young, R. Lewyckyj and Y. Takane. (1986). The ALSCAL Procedure. SUGI Supplemental Library User's Guide. Version 5 Edition. SAS Institute Inc.

Chapter 9

USING THE GAMMA DISTRIBUTION TO FIT FECUNDITY CURVES FOR APPLICATION IN ANDALUSIA (SPAIN)

F. ABAD-MONTES

Dpto. Estadística e Investigación Operativa, Universidad de Granada C/Fuentenueva,s/n, Granada, España

M.D. HUETE-MORALES

Dpto. Estadística e Investigación Operativa, Universidad de Granada C/Fuentenueva,s/n, Granada, España

M. VARGAS-JIMÉNEZ

Dpto. Estadística e Investigación Operativa, Universidad de Granada, C/Fuentenueva,s/n, Granada, España

Analysis of the evolution of specific fecundity rates, by the age of the mother, i.e. fecundity curves, and their modelling, is of vital importance when we seek to obtain projections or forecasts of the behaviour of this demographic phenomenon. Indeed, on some occasions these estimates do not need to be reasonable from the populational standpoint, but may have the goal of establishing hypothetical scenarios. The present study includes an analysis of the observed data for total births (without taking into account the order of birth) by age and by female population. These data, for the period 1975–2001, were provided by the Statistical Institute of Andalusia (IEA) and were used to construct synthetic fecundity indicators, which are the most basic and the most effective means of accounting for the global behaviour pattern of the phenomenon within a given period. Subsequently, the observed fecundity curves were fitted using a Gamma-type distribution. This distribution is one of the most commonly used, for two main reasons: it provides very good quality fits, and the parameters of the distribution are identified perfectly with the indicators of fecundity. Finally, various behaviour hypotheses are proposed, on the basis of the information obtained during the period of analysis.

1. Data utilized and basic indicators

In order to address the demographic phenomenon we are concerned with, we must first obtain a series of fecundity rates ranked by the mother's age, this series being known as the Fecundity Curve. The following data were provided by the Statistical Institute of Andalusia (IEA): number of births by mother's age and by age of the female population, on 1 January of each year being considered, which in the case of the present study was 1975–2001, within the area comprising the Autonomous Community of Andalusia. The study was carried out for a population of women of fertile age, this being taken as ages 15 to 49. With this information, we calculated the specific fecundity rates for each age (x) and each year (t), these rates being denoted by f_x^t:

$$f_x^t = \frac{N_x^t}{\dfrac{P_x^{1/1/t} + P_x^{1/1/t+1}}{2}} \tag{1}$$

where N_x^t is the number of births to mothers who have passed their 'x' birthday during year 't', and $P_x^{1/1/t}$ is the female population having passed their 'x' birthday by 1 January in year 't'. These rates, for some of the years in question, are represented as follows:

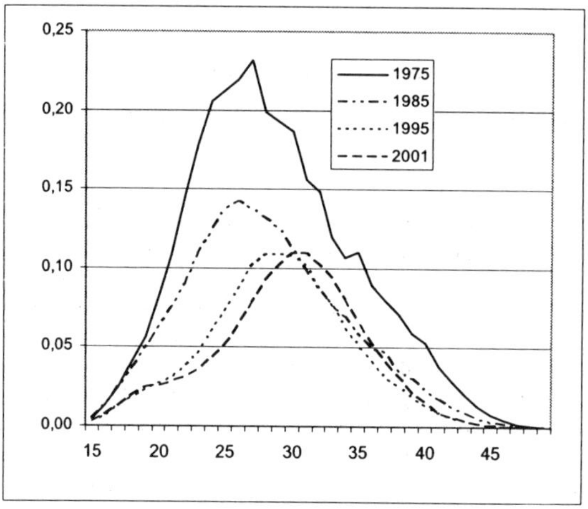

Figure 1. Fecundity curves in Andalusia

It is very apparent that in little more than a quarter of a century the pattern of fecundity in Andalusia has varied spectacularly. In 1975, fecundity rates were very high for almost all the ages, which suggests that the number of births was also high. These high rates were mainly due to the fact that families began to have children at a fairly young age and went on to have a lot of them; this explains why fecundity rates were so high at the end of the fertile period. This situation did not last, however, and the above figure shows that by 1985 the fecundity rates had fallen significantly. Subsequently, they continued to fall, though less dramatically. Nevertheless, it can be seen that the bell shape of the fecundity curve was distorted, with the mode of the distribution shifting to the right (as a result of the age of first pregnancy being delayed) and the appearance of a "second mode", which reflects the births that occur to very young mothers, normally unmarried and of children who were often unplanned.

Let us now define and construct the most commonly used indicators of fecundity. First, we obtain the Synthetic Fecundity Index (SFI) which describes the mean number of children per woman of fertile age:

$$SFI^t = \sum_{x=15}^{49} f_x^t \tag{2}$$

Other relevant indicators include the Mean Age at Maternity (MAM), which describes whether the age of maternity is rising or falling, and the Variance in the Age at Maternity (VAM), which provides a measure of the variability of the occurrence of births, i.e. whether these occur at widely-spaced ages or are closely grouped around the mean age:

$$MAM^t = \frac{\sum_{x=15}^{49}(x+0{,}5)f_x^t}{\sum_{x=15}^{49} f_x^t} \tag{3}$$

$$\sigma^{2t} = VAM^t = \frac{\sum_{x=15}^{49}\left[(x+0{,}5)-MAM\right]^2 f_x^t}{\sum_{x=15}^{49} f_x^t} \tag{4}$$

Table 1 shows the application of the above expressions to the available information. The pattern of this series of indices might be more apparent in graphical form:

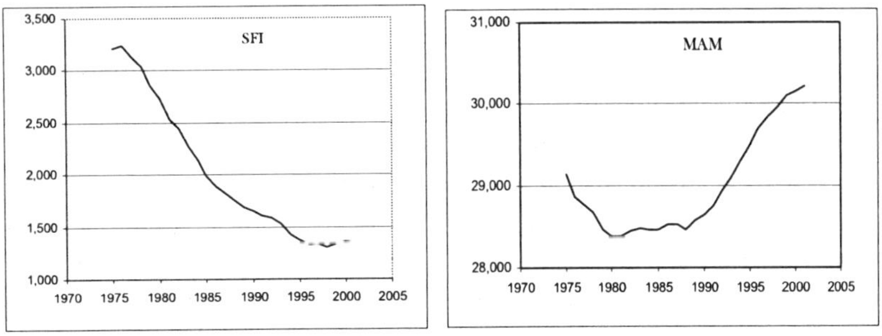

Figure 2. Variation in SFI and MAM in Andalusia

Table 1. Variation of SFI, MAM and VAM in Andalusia

Year	SFI	MAM	VAM
1975	3.212	29.138	35.882
1976	3.238	28.873	35.225
1977	3.132	28.769	35.685
1978	3.041	28.675	35.740
1979	2.861	28.469	35.905
1980	2.739	28.387	36.236
1981	2.535	28.388	35.787
1982	2.444	28.453	35.326
1983	2.275	28.484	34.918
1984	2.140	28.472	34.823
1985	1.990	28.470	34.324
1986	1.891	28.529	34.015
1987	1.819	28.525	33.141
1988	1.760	28.471	32.322
1989	1.689	28.576	31.556
1990	1.656	28.636	30.087
1991	1.612	28.758	29.807
1992	1.581	28.936	29.087
1993	1.527	29.095	28.262
1994	1.426	29.305	28.196
1995	1.375	29.493	27.931
1996	1.329	29.704	27.518
1997	1.336	29.843	27.908
1998	1.303	29.961	28.088
1999	1.335	30.099	28.527
2000	1.358	30.157	29.011
2001	1.354	30.209	29.283

The Synthetic Fecundity Index and that of the mean age of maternity reveal a very different behaviour pattern; the former has fallen gradually over the years, from 3.2 children per woman in 1975 to 1.3 in the year 2001. With respect to the mean age of maternity, the graph might be considered to present a distorted view of reality, since although the mean age seems to fall in the initial years, then stabilise and then rise from the late 1980s onwards, we must take into account the very high values recorded at the beginning of this period. This latter fact was due to the very long period of fecundity commonly presented then, with mothers having a large number of children; thus, the mean age of maternity was higher than that of mothers today.

This situation is reflected in the Index of the Variance; in the initial years of the study, the variance was very high, and so births were not concentrated around the mean, but widely distributed throughout the fertile life of the mothers:

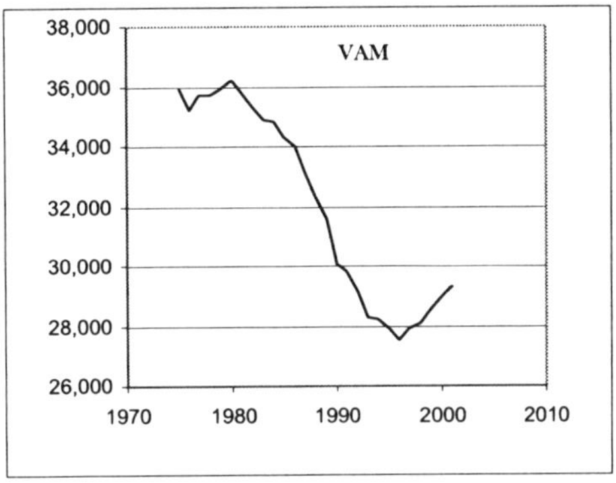

Figure 3. Variation in VAM in Andalusia

It should be noted that in very recent years there has been a moderate rise in the SFI (which shows that women in Andalusia are starting to have more children), a levelling off in the rise in the mean age of maternity and a rise in the variance (partly due to the "second mode", referred to above, in the fecundity curves).

2. Fitting and modelling the fecundity curves

The series of specific rates of fecundity by age for each year of the observed series can be fitted by means of various distributions, including the Hadwiger, Lognormal, Mirás and Beta functions and, of course, the one that is most often used, the Gamma (or Pearson type III) distribution, because of its extraordinary advantages.

These advantages include its ease of application, the very acceptable fits it produces and the fact that its parameters are identified with the above-listed indicators (SFI, MAM, VAM), i.e. it depends on them.

The following expression is used to fit the fecundity curve:

$$F(y) = \frac{ab^c y^{c-1} \exp\{-by\}}{\Gamma(c)} \tag{5}$$

where y is the class mark of the interval considered less the minimum fertile age, i.e. $y = (x + 0{,}5) - 15$ and $\Gamma(c)$ is the gamma function. The parameters a, b and c of F(y) are related to the fertility indicators as follows:

$$a = SFI \quad b = \frac{\bar{y}}{\sigma^2} \quad c = \frac{\bar{y}^2}{\sigma^2} \tag{6}$$

Thus, by fitting the above to the series of rates per year, we obtain:

Table 2. Fertility indicators in Andalusia

Year	a	b	C
1975	3.212	0.39402	5.57054
1976	3.238	0.39384	5.46361
1977	3.132	0.38584	5.31246
1978	3.041	0.38263	5.23262
1979	2.861	0.37512	5.05241
1980	2.739	0.36942	4.94530
1981	2.535	0.37411	5.00861
1982	2.444	0.38081	5.12291
1983	2.275	0.38615	5.20674
1984	2.140	0.38687	5.21184
1985	1.990	0.39243	5.28599
1986	1.891	0.39774	5.38110
1987	1.819	0.40812	5.51987
1988	1.760	0.41678	5.61443
1989	1.689	0.43020	5.84028
1990	1.656	0.45324	6.18063
1991	1.612	0.46157	6.35014
1992	1.581	0.47910	6.67658
1993	1.527	0.49871	7.02920
1994	1.426	0.50733	7.25726
1995	1.375	0.51888	7.52001
1996	1.329	0.53433	7.85663
1997	1.336	0.53184	7.89414
1998	1.303	0.53263	7.96850
1999	1.335	0.52930	7.99202
2000	1.358	0.52244	7.91845
2001	1.354	0.51938	7.89913

Below, we illustrate some of the fits that were made:

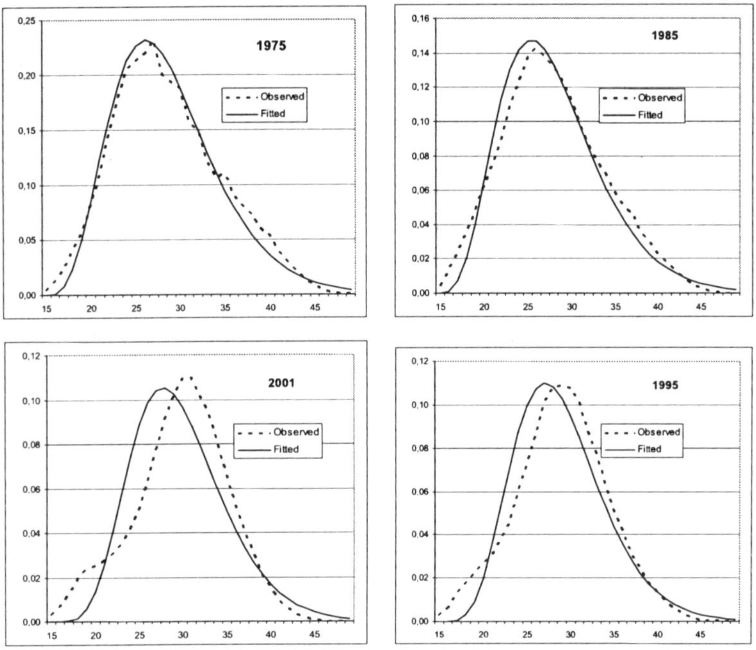

Figure 4. Specifics rates in Andalusia

These figures show that the Gamma fit for the fecundity curves is considerably better in the initial years; the "second mode" that appears for the early ages, in the latter years of the series, is worse fitted, as in all of them the curve is shifted to the left.

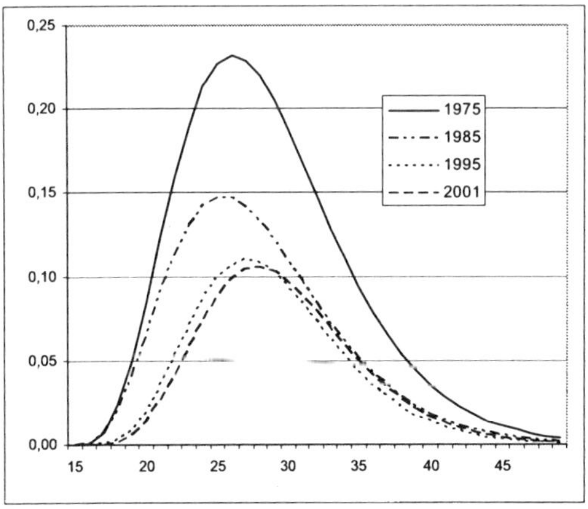

Figure 5. Fitted fecundity curves for Andalusia

3. Use of the fit in forecasts of fecundity curves

Let us now make a forecast (or rather, a simulation) of the fecundity curve that might be recorded for Andalusia for the coming years. To establish reasonable hypotheses of future behaviour, it would be necessary to perform a more exhaustive study of the current characteristics of fecundity in this region, as regards fecundity by order of birth, within and outside the marriage, by foreigners, and many other parameters. However, this is not the aim of the present analysis; rather, we seek to perform simulations of the fecundity rates under various more or less plausible scenarios of behaviour. Therefore, we shall limit ourselves to establishing different hypotheses about the synthetic parameters of fecundity. Let us examine the trend in the series of indicators for recent years:

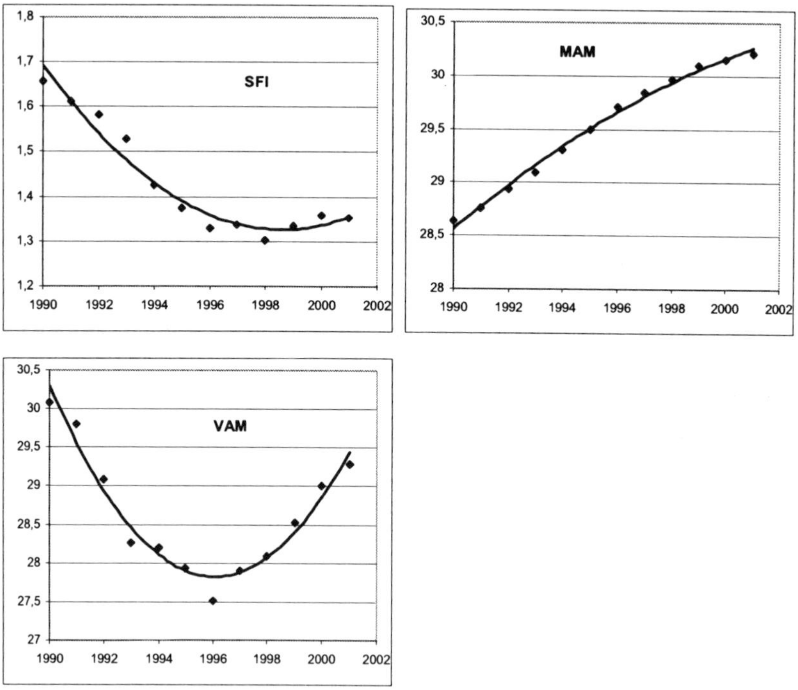

Figure 6. Indicators for recent year

A clear pattern can be observed in all the series. The SFI, although it has fallen, seems to have recovered in the last few years; the MAM is also increasing, albeit slowly (which might be a consequence of the fact that the SFI is improving); and what is most dramatic is the recovery of the variance (which could indicate that women in Andalusia are having children at more widely spaced intervals, and perhaps too that the number of children born in higher orders of birth is greater). Taking all this into account, we assume the following values:

$$1^{st}\ Hypothesis \begin{cases} SFI = 1{,}4 \\ MAM = 30{,}4 \\ VAM = 29{,}8 \end{cases}$$

$$2^{nd}\ Hypothesis \begin{cases} SFI = 1{,}6 \\ MAM = 31 \\ VAM = 30 \end{cases}$$

$$3^{rd}\ Hypothesis \begin{cases} SFI = 1{,}3 \\ MAM = 29 \\ VAM = 28 \end{cases}$$

These hypotheses would correspond, respectively, to: 1) a slight improvement in fecundity rates in Andalusia; 2) a markedly higher number of children being born to each woman, with women having children at wider age intervals; 3) fewer children born to each woman and an advance in the mean age of maternity. Let us examine a graphic representation of these three hypotheses, compared to the observed data for 2001:

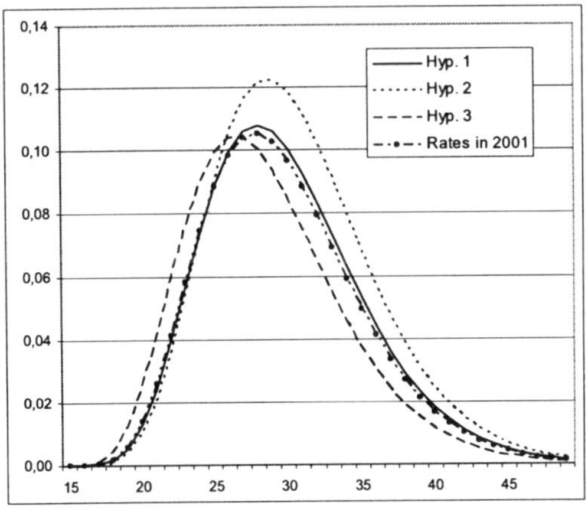

Figure 7. Forecast fecundity curves

Table 3. Fertility indicators in Andalusia (hypotheses 1, 2 and 3)

	SFI	MAM	VAM	a	b	c
Hypothesis 1	1.4	30.4	29.8	1.400	0.51678	7.95839
Hypothesis 2	1.6	31.0	30.0	1.600	0.53333	8.53333
Hypothesis 3	1.3	29.0	28.0	1.300	0.50000	7.00000

4. Influence of the Index of Generational Replacement

The Index of Generational Replacement is the number of children per woman that is necessary so that the study population can replace itself; this indicator would correspond to an SFI of 2.1 children per woman.

Let us implement a simulation exercise in which in the coming years each woman in the population has 2.1 children. According to the data for the current situation in Andalusia, in order to reach this level there would first have to be an increase in the variance of the age of maternity, as births would be more widely dispersed with respect to age. Therefore, let us assume a value of 30. As concerns the age at maternity, this too would have to rise (the current trend for women is to have their children at later ages), and so we would take a mean age of 30.5 years.

The result of the projection of the fecundity curve, under the above assumptions for the fecundity characteristics, is shown below.

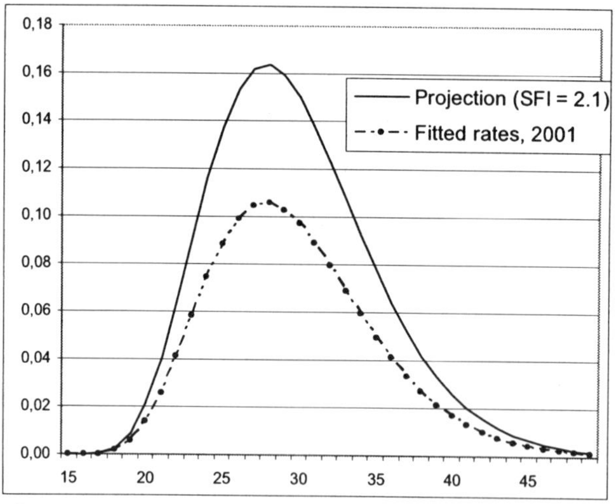

Figure 8. Fecundity curves

5. Conclusions

The Gamma distribution is a powerful tool for the analysis and subsequent projection of fecundity curves for a given zone, and this distribution is the most widely used by demographers and other researchers in the field. The results obtained reveal its clarity and suitability for modelling fecundity patterns and for carrying out simulations or predictions of future patterns, largely because its parameters depend on synthetic indicators of fecundity.

References

1. Arroyo, A. (coordinator), Hernández, J., Romero, J., Viciana, F. and Zoido, F. (2004). Tendencias demográficas durante el siglo XX en España. INE. Madrid.
2. Brass, W. (1971). Seminario sobre modelos para medir variables demográficas (Fecundidad y mortalidad). CELADE. S. José de Costa Rica.
3. Brass, W. (1974). Métodos para estimar la fecundidad y la mortalidad en poblaciones con datos limitados. CELADE. Santiago de Chile.
4. I.E.A. (1999). Un siglo de demografía en Andalucía. La población desde 1900. Sevilla.
5. Leridon, H. and Toulemon, L. (1997). Démographie. Economica. Paris.
6. Pressat, R. (1995). Eléments de démographie mathématique. AIDELF. París.

Chapter 10

CLASSES OF BIVARIATE DISTRIBUTIONS WITH NORMAL AND LOGNORMAL CONDITIONALS: A BRIEF REVISION*

J.M. SARABIA

Department of Economics, University of Cantabria
Avda. de los Castros s/n. Santander, 39005, Spain

E. CASTILLO

Dept. of Applied Mathematics and Computational Sciences
University of Cantabria
. Avda. de los Castros s/n. Santander, 39005, Spain

M. PASCUAL

Department of Economics, University of Cantabria
Avda. de los Castros s/n. Santander, 39005, Spain

M. SARABIA

Department of Business Administration, University of Cantabria
Avda. de los Castros s/n. Santander, 39005, Spain

The present paper is a brief survey of the classes of bivariate distributions with normal and lognormal conditionals. Basic properties including conditional moments, marginal distributions, characterizations, parameterizations, dependence and modality are revised. Estimation and applications of these models are studied. Finally, some extensions of the bivariate conditional normal model are reviewed.

1. Introduction and motivation

Let (X,Y) be a bivariate random variable with joint probability density function (pdf) $f(x,y)$. It is well known that the couple of marginal distributions does not determine the bivariate distribution. For example, a bivariate distribution with normal marginals need not to be a classical bivariate normal density. Probably, the simplest example is

*The authors thank to Ministerio de Educación y Ciencia (project SEJ2004-02810) for partial support of this work.

$$f(x,y) = \frac{1}{\pi} \exp\left[-\left(x^2 + y^2\right)/2\right] I(xy > 0)$$

where $I(\cdot)$ is the indicator function. However, the conditional distributions functions uniquely determine a joint density function [1].

The classical bivariate normal distribution has both marginal and conditional normal distributions. A natural question then arises: Must a bivariate distribution with normal conditionals be a classical bivariate normal distribution? The answer is negative, and then, another question arises: What is this class of distributions?

We answer this question, studying and reviewing the class of bivariate distributions with normal conditionals and next we will study the class of distribution with lognormal conditionals. The present paper surveys bivariate distributions with normal and lognormal conditionals.

2. Bivariate distribution with normal conditionals

Assume that (X, Y) is a random vector that has a joint density. The marginal, conditional and joint densities are denoted by $f_X(x)$, $f_Y(y)$, $f_{X|Y}(x \mid y)$, $f_{Y|X}(y \mid x)$ and $f(x,y)$. We are interested in obtaining the most general bivariate random variable whose conditional distributions are normal,

$$X \mid Y = y \sim N\left(\mu_1(y), \sigma_1^2(y)\right), \tag{1}$$

$$Y \mid X = x \sim N\left(\mu_2(x), \sigma_2^2(x)\right), \tag{2}$$

that is

$$f_{X|Y}(x \mid y) = \frac{1}{\sigma_1(y)\sqrt{2\pi}} \exp\left[-\frac{1}{2}\left(\frac{x - \mu_1(y)}{\sigma_1(y)}\right)^2\right], \tag{3}$$

$$f_{Y|X}(y \mid x) = \frac{1}{\sigma_2(x)\sqrt{2\pi}} \exp\left[-\frac{1}{2}\left(\frac{y - \mu_2(x)}{\sigma_2(x)}\right)^2\right], \tag{4}$$

where $u_i(u): \Re \to \Re$, $i = 1,2$ and $\sigma_i(u): \Re \to \Re^+$, $i = 1,2$ are unknown functions.

This bivariate distribution was obtained by Castillo and Galambos [2,3]. Bhattacharyya [4] obtained the same expression solving another different problem.

If we write the joint density as product of marginals and conditionals we obtain the functional equation:

$$\frac{f_X(x)}{\sigma_2(x)\sqrt{2\pi}}\exp\left[-\frac{1}{2}\left(\frac{y-\mu_2(x)}{\sigma_2(x)}\right)^2\right]=\frac{f_Y(y)}{\sigma_1(y)\sqrt{2\pi}}\exp\left[-\frac{1}{2}\left(\frac{x-\mu_1(y)}{\sigma_1(y)}\right)^2\right].$$

This is a functional equation with 6 different unknown functions. There are two ways for solving this functional equation: using general methods of functional equations or using standard calculus techniques. For this kind of equations, functional equation's methods have been used widely by Arnold, Castillo and Sarabia [5]. In this particular case, it is possible to use standard calculus. Taking logarithms we get:

$$\begin{cases}\log f(x,y)=\log\left(\dfrac{f_Y(y)}{\sigma_1(y)\sqrt{2\pi}}\right)-\dfrac{1}{2\sigma_1(y)^2}(x-\mu_1(y))^2,\\[2ex] \log f(x,y)=\log\left(\dfrac{f_X(x)}{\sigma_2(x)\sqrt{2\pi}}\right)-\dfrac{1}{2\sigma_2(x)^2}(y-\mu_2(x))^2.\end{cases}$$

We write:

$$\begin{cases}\log f(x,y)=a_1(y)+b_1(y)x+c_1(y)x^2,\\ \log f(x,y)=\widetilde{a}_1(x)+\widetilde{b}_1(x)y+\widetilde{c}_1(x)y^2.\end{cases}$$

Now, if we assume differentiability for $f(x,y)$,

$$\begin{cases}\dfrac{\partial^2\log f(x,y)}{\partial x^2} &= 2c_1(y),\ \forall y\\[2ex] \dfrac{\partial^2\log f(x,y)}{\partial x^2} &= \widetilde{a}_1''(x)+\widetilde{b}_1''(x)y+\widetilde{c}_1''(x)y^2,\ \forall y.\end{cases}$$

In consequence, the functions $c_1(y)$ and $\widetilde{a}_1(y)$, $\widetilde{b}_1(y)$, $\widetilde{c}_1(y)$ must be polynomials of degree 2. Finally, computing $\left(\partial^2/\partial y^2\right)\log f(x,y)$, we conclude that $a_1(y)$ and $b_1(y)$ are polynomial of degree two. Finally, the joint pdf $f(x,y)$ must be of this form,

$$f(x,y)=\exp\left\{m_{00}+m_{10}x+m_{01}y+m_{20}x^2+m_{02}y^2+m_{11}xy+m_{12}xy^2+m_{21}x^2y+m_{22}x^2y^2\right\}. \tag{5}$$

2.1. *Conditional moments and marginal distributions*

From general expression (5) and by identification with (3) and (4) we obtain,

$$\log\left(\frac{f_Y(y)}{\sigma_1(y)\sqrt{2\pi}}\right)-\frac{\mu_1^2(y)}{2\sigma_1^2(y)} = m_{00}+m_{01}y+m_{02}y^2,$$

$$\frac{\mu_1(y)}{\sigma_1^2(y)} = m_{10}+m_{11}y+m_{12}y^2,$$

$$-\frac{1}{2\sigma_1^2(y)} = m_{20}+m_{21}y+m_{22}y^2,$$

which leads to

$$E(X|Y=y) = \mu_1(y) = -\frac{m_{12}y^2+m_{11}y+m_{10}}{2(m_{22}y^2+m_{21}y+m_{20})}, \tag{6}$$

$$\operatorname{Var}(X|Y=y) = \sigma_1^2(y) = \frac{-1}{2(m_{22}y^2+m_{21}y+m_{20})}, \tag{7}$$

$$E(Y|X=x) = \mu_2(x) = -\frac{m_{21}x^2+m_{11}x+m_{01}}{2(m_{22}x^2+m_{12}x+m_{02})}, \tag{8}$$

$$\operatorname{Var}(Y|X=x) = \sigma_1^2(x) = \frac{-1}{2(m_{22}x^2+m_{12}x+m_{02})}. \tag{9}$$

The marginal densities are given by:

$$f_X(x)=\left(-2(m_{22}x^2+m_{12}x+m_{02})\right)^{-\frac{1}{2}}\times \exp\left\{\frac{1}{2}\left[2(m_{20}x^2+m_{10}x+m_{00})-\frac{(m_{21}x^2+m_{11}x+m_{01})^2}{2(m_{22}x^2+m_{12}x+m_{02})}\right]\right\}, \tag{10}$$

$$f_Y(x)=\left(-2(m_{22}y^2+m_{21}y+m_{20})\right)^{-\frac{1}{2}}\times \exp\left\{\frac{1}{2}\left[2(m_{02}y^2+2m_{01}y+m_{00})-\frac{(m_{12}y^2+m_{11}y+m_{10})^2}{2(m_{22}y^2+m_{21}y+m_{20})}\right]\right\}. \tag{11}$$

The joint pdf can be written in the alternative form (with the notation used for more general exponential families):

$$f(x,y)=\exp\left\{(1,x,x^2)\begin{pmatrix} m_{00} & m_{01} & m_{02}\\ m_{10} & m_{11} & m_{12}\\ m_{20} & m_{21} & m_{22}\end{pmatrix}\begin{pmatrix}1\\ y\\ y^2\end{pmatrix}\right\}. \tag{12}$$

It remains only to determine appropriate conditions on the constants m_{ij}; $i, j = 0, 1, 2$ in (12) to ensure the integrability of those marginals. The constant m_{00} will be a function of the other parameters.

2.2. *Properties of the normal conditionals distribution*

The normal conditionals distribution has joint density of the form (12) where the m_{ij} constants satisfy one of the two sets of conditions

(a) $m_{22} = m_{12} = m_{21} = 0$; $m_{20} < 0$; $m_{02} < 0$; $m_{11}^2 < 4m_{02}m_{20}$.

(b) $m_{22} < 0$; $4m_{22}m_{02} > m_{12}^2$; $4m_{20}m_{22} > m_{21}^2$.

Models satisfying conditions (a), are the classical bivariate normal models with:

- Normal marginals,
- Normal conditionals,
- Linear regressions and constant conditional variances.

More interesting are the models satisfying conditions (b). These models have:

- Normal conditional distributions,
- Non-normal marginal densities (see (10) and (11)),
- The regression functions are either constant or non-linear given by (6) and (8). Each regression function is bounded (in contrast with the classical bivariate normal model).
- The conditional variance functions are also bounded and non constant. They are given by (7) and (9).

What if we require normal conditionals and independent marginals? Referring to (12) the requirement of independence translates to the following functional equation

$$\left(1, x, x^2\right)\begin{pmatrix} m_{00} & m_{01} & m_{02} \\ m_{10} & m_{11} & m_{12} \\ m_{20} & m_{21} & m_{22} \end{pmatrix}\begin{pmatrix} 1 \\ y \\ y^2 \end{pmatrix} = r(x) + s(y), . \tag{13}$$

Its solution eventually leads us to

$$m_{11} = m_{21} = m_{12} = m_{22} = 0$$

which is the independence model. This result shows that independence is only possible within the classical bivariate normal model. As consequences of the above discussion, Castillo and Galambos [3] derive the following interesting conditional characterizations of the classical bivariate normal distribution.

Theorem $f(x,y)$ is a classical bivariate normal density if and only if all conditional distributions, both of *X* given *Y* and of *Y* given *X*, are normal and any one of the following properties hold

(i) $\sigma_2^2(x) = \text{Var}\left(Y \mid X = x\right)$ or $\sigma_1^2(y) = \text{Var}\left(X \mid Y = y\right)$ is constant

(ii) $\lim_{y\to\infty} y^2\sigma_1^2(y) = \infty$ or $\lim_{x\to\infty} x^2\sigma_2^2(x) = \infty$

(iii) $\underline{\lim}_{y\to\infty}\sigma_1(y) \neq 0$ or $\underline{\lim}_{x\to\infty}\sigma_2(x) \neq 0$

(iv) $E(Y \mid X = x)$ or $E(X \mid Y = y)$ is linear and non-constant.

Proof: Direct, using the general expression for $f(x,y)$.

Other characterizations of the classical bivariate normal distribution by conditional properties have been proposed by Ahsanullah [6], Arnold, Castillo and Sarabia [7,8], Bischoff [9,10,11], Bischoff and Fieger [12] and Nguyen, Rempala and Wesolowski [13].

2.3. *Convenient parameterizations*

Expression (5) depends on 8 parameters, and the normalizing constant is not available in a close form. From a practical point of view it is convenient to provide some simpler models or some convenient parameterization. In this way Gelman and Meng [14] proposed a simple parameterization. If in (12) we make location and scale transformations in each variable we get:

$$f(x,y) \propto \exp\left\{-\left(\alpha x^2y^2 + x^2 + y^2 + \beta xy + \gamma x + \delta y\right)\right\}, \tag{14}$$

where α, β, γ, and δ are the new parameters which are functions of the old m_{ij} parameters. In this parameterization, the conditional distributions are

$$X \mid Y = y \sim N\left(-\frac{\beta y + \gamma}{2(\alpha y^2 + 1)}, \frac{1}{2(\alpha y^2 + 1)}\right),$$

$$Y \mid X = x \sim N\left(-\frac{\beta x + \delta}{2(\alpha x^2 + 1)}, \frac{1}{2(\alpha x^2 + 1)}\right).$$

The only constraints for this parameterization are:

$$\alpha \geq 0 \text{ and if } \alpha = 0 \text{ then } \left|\beta\right| < 2. \tag{15}$$

An advantage of this Gelman and Meng parameterization is that multimodality can be studied easily.

Other important parameterization was proposed by Sarabia [15]. This author proposed the choice $\mu_i(u) = \mu_i, \quad i = 1, 2$ for obtaining the joint density

$$f(x,y;\mu,\sigma,c)=\frac{k(c)}{2\pi\sigma_1\sigma_2}\exp\left\{-\frac{1}{2}\left(z_1^2+z_2^2+cz_1^2z_2^2\right)\right\},\quad x,y>0 \tag{16}$$

Where $z_1=(x-\mu_1)/\sigma_1$, $z_2=(y-\mu_2)/\sigma_2$ and $c\geq 0$. In this case the conditional distributions are given by

$$X\mid Y=y\sim N\left(\mu_1,\frac{\sigma_1^2}{1+c(y-\mu_2)^2/\sigma_2^2}\right),$$

$$Y\mid X=x\sim N\left(\mu_2,\frac{\sigma_2^2}{1+c(x-\mu_1)^2/\sigma_1^2}\right),$$

and the normalizing constant by

$$k(c)=\frac{\sqrt{2c}}{U(1/2,1,1/2c)},$$

where $U(a,b,z)$ represents the confluent hypergeometric function $(a,z>0)$

$$U(a,b,z)=\frac{1}{\Gamma(a)}\int_0^\infty e^{-tz}t^{a-1}(1+t)^{b-a-1}\mathrm{d}t\,.$$

2.4. *How many modes?*

Gelman and Meng [14] gave an example of a distribution of this type with two modes. The conditional mode curves (which correspond to the conditional mean curves (6) and (8)) can intersect in more than one point. The general problem was solved by Arnold, Castillo, Sarabia and Gonzalez-Vega [16]. Since in this model, modes are at the intersection of regression lines, the coordinates of the modes satisfy the following system of equations

$$\begin{aligned} x&=-\frac{\beta y+\gamma}{2(\alpha y^2+1)} \\ y&=-\frac{\beta x+\delta}{2(\alpha x^2+1)}. \end{aligned} \tag{17}$$

Substituting the first into the second we get

$$4\alpha^2y^5+2\alpha^2\delta y^4+8\alpha y^3+\alpha(4\delta+\beta\gamma)y^2+(4-\beta^2+\alpha\gamma^2)y+2\delta-\beta\gamma=0,$$

which is a polynomial of degree five. When this polynomial has a unique real root, the density is unimodal, if it has three distinct real roots (two modes and a saddle point) the density is bimodal, and with 5 distinct real roots (three relative maxima and two saddle points) we have 3 modes. For example, in the symmetric case $\delta = \gamma$, we have 3 real roots and consequently $f(x,y)$ will be bimodal if and only if

$$\alpha\delta^2 > 8(2-\beta).$$

2.5. *Dependence*

For this model the usual correlation coefficient is not limited. Other alternative non-scalar dependence measure is the local dependence function [17,18] defined by

$$\gamma(x,y) = \frac{\partial^2 \log f(x,y)}{\partial x\, \partial y}, \tag{18}$$

which gives more detailed information about the dependence. In this case, the local dependence function is

$$\gamma(x,y) = \frac{\partial^2 \log f(x,y)}{\partial x\, \partial y} = m_{11} + 2m_{21}x + 2m_{12}y + 4m_{22}xy.$$

An interpretation of this function is possible: random variables X and Y are positively associated in the first and third quadrants and negatively associated in the second and fourth which supposes non-linear dependences in the model.

3. Interesting properties and applications of models with conditional specification

The properties of the model with normal conditionals (some of them unexpected properties) appear in another model with conditional specification. We enumerate some of these properties of these conditional models.

- Models with conditional specification use to depend on a large number of parameters.
- They include as particular cases the independence case and well known classical models.
- Its dependence structure is richer than that for the usual models (local and global).
- Sometimes they present multimodality.
- Characterizations of some classical models can be obtained based on these models.
- In the most limited case, if we begin with a one-parameter family, the dependence structure can be limited. However the marginal

distributions are wider than the usual models. They present overdispersion.

- The resulting densities are easy to simulate using Gibbs sampler techniques, indeed they are tailor made for such simulation.

4. Bivariate distributions with lognormal conditionals

In this section the class of bivariate distributions with log-normal conditionals is reviewed. This model has been recently proposed by Sarabia, Castillo, Pascual and Sarabia [19] and has important applications in the study of bivariate income distributions. We work with the triparametric version of the lognormal distribution that we will denote by $X \sim LN(\mu, \sigma, \delta)$ and has pdf,

$$f(x;\mu,\sigma) = \frac{1}{(x-\delta)\sigma\sqrt{2\pi}} \exp\left\{-\frac{1}{2}\left(\frac{\log(x-\delta)-\mu}{\sigma}\right)^2\right\}, \quad x > \delta$$

with $\mu \in \Re$ y $\sigma > 0$. Then, we are interested in the more general random variable (X, Y) satisfying,

$$X \mid Y = y \sim LN\left(\delta_1, \mu_1(y), \sigma_1(y)\right), \tag{19}$$

$$Y \mid X = x \sim LN\left(\delta_2, \mu_2(x), \sigma_2(x)\right). \tag{20}$$

If conditions (19) and (20) are satisfied, the joint probability density function takes the form,

$$f(x,y;\delta,m) = (x-\delta_1)^{-1}(y-\delta_2)^{-1}\exp\left\{-u_{\delta_1}(x)^{\mathrm{T}} M u_{\delta 2}(y)\right\} \tag{21}$$

where $u_{\delta i}(\cdot)$ denotes the vector

$$u_{\delta_i}(z) = \left(1, \log(z-\delta_i), \left[\log(z-\delta_i)\right]^2\right)^{\mathrm{T}}, \quad i = 1, 2,$$

and $M = \{m_{ij}\}$ is a 3×3 parameters matrix. The parameters $\{m_{ij}\}$ must be chosen such that (21) to be integrable. Expanding formula (21) we obtain the formula:

$$f(x,y;\delta,m) = \left[(x-\delta_1)(y-\delta_2)\right]^{-1}\exp\left\{-\left[m_{00} + u(z_1,z_2) + v(z_1,z_2)\right]\right\} \tag{22}$$

where

$$\begin{aligned} u(x,y) &= m_{10}x + m_{20}x^2 + m_{01}y + m_{02}y^2 + m_{11}xy, \\ v(x,y) &= m_{12}xy^2 + m_{21}x^2y + m_{22}x^2y^2 \end{aligned}$$

and $z_1 = \log(x-\delta_1)$, $z_2 = \log(x-\delta_2)$. The function $u(\cdot,\cdot)$ contains the terms that appear in the classical model and the function $v(\cdot,\cdot)$ contains new terms that appear in these conditional models. The conditional parameters $\mu_i(u)$ and $\sigma_i(u)$, $i = 1, 2$ are:

$$\mu_1(y) = -\frac{m_{12}z_2^2 + m_{11}z_2 + m_{10}}{2(m_{22}z_2^2 + m_{21}z_2 + m_{20})}, \tag{23}$$

$$\mu_2(x) = -\frac{m_{21}z_1^2 + m_{11}z_1 + m_{01}}{2(m_{22}z_1^2 + m_{12}z_1 + m_{02})}, \tag{24}$$

and

$$\sigma_1^2(y) = \frac{1}{2(m_{22}z_2^2 + m_{21}z_2 + m_{20})}, \tag{25}$$

$$\sigma_2^2(x) = \frac{1}{2(m_{22}z_1^2 + m_{12}z_1 + m_{02})}. \tag{26}$$

4.1. *General properties*

The constant $\exp(m_{00})$ is the normalizing constant and it is a function of the rest of parameters. In order to have a genuine joint pdf, sufficient conditions for integrability of (22) are that the parameters satisfy one of the following two sets of conditions:

$$m_{12} = m_{21} = m_{22} = 0, \quad m_{02} > 0, \quad m_{20} > 0, \quad m_{11}^2 < 4m_{02}m_{20}, \tag{27}$$

$$m_{22} > 0, \quad m_{12}^2 < 4m_{22}m_{02}, \quad m_{21}^2 < 4m_{22}m_{20}. \tag{28}$$

If (27) is satisfied, we obtain the classical bivariate lognormal distribution. If (28) satisfies, we find a new class of distributions. The marginal distributions are given by $(x > \delta_1)$:

$$f_X(x;\delta_1,m) = \frac{\exp\left\{-\left[(m_{20}z_1^2 + m_{10}z_1 + m_{00}) + \frac{(m_{21}z_1^2 + m_{11}z_1 + m_{01})^2}{4(m_{22}z_1^2 + m_{12}z_1 + m_{02})}\right]\right\}}{\sqrt{(m_{22}z_1^2 + m_{12}z_1 + m_{020})/2\pi}} \tag{29}$$

and $(y > \delta_2)$

$$f_Y(y;\delta_2,m)=\frac{\exp\left\{-\left[\left(m_{02}z_2^2+m_{01}z_2+m_{00}\right)+\frac{\left(m_{12}z_2^2+m_{11}z_2+m_{10}\right)^2}{4\left(m_{22}z_2^2+m_{21}z_2+m_{20}\right)}\right]\right\}}{\sqrt{\left(m_{22}z_2^2+m_{21}z_2+m_{20}\right)/2\pi}}. \tag{30}$$

Note that (29) and (30) are not lognormal distributions if conditions (28) hold. These marginals depend on all eight parameters and then present a high flexibility. The conditional moments of (22) are $(r=1,2,...)$:

$$E\left[(X-\delta_1)^r \mid Y\right]=\exp\{r\mu_1(Y)+r^2\sigma_1^2(Y)/2\}, \tag{31}$$

$$E\left[(Y-\delta_2)^r \mid X\right]=\exp\{r\mu_2(X)+r^2\sigma_2^2(X)/2\}, \tag{32}$$

where $\mu_i(U)$ and $\sigma_i^2(U)$, $i=1,2$ are given in (23) to (25). Combining (31)-(32) with (29)-(30) the moments of the marginal marginal distributions as well as the correlation coefficient can be obtained. The usual correlation coefficient is not obviously limited and the local dependence function (18) is

$$\gamma(x,y)=-\frac{m_{11}+2m_{21}\log(x)+2m_{12}\log(y)+4m_{22}\log(x)\log(y)}{xy}.$$

5. Estimation

For this kind of conditional models, several estimation strategies have been proposed by Arnold, Castillo and Sarabia [5]. Here we pay attention in techniques based on the likelihood. The family of densities (5) is a member of the exponential family with natural sufficient statistics:

$$\left(\sum x_i,\sum y_i,\sum x_i^2,\sum y_i^2,\sum x_iy_i,\sum x_iy_i^2\sum x_i^2y_i,\sum x_i^2y_i^2\right), \tag{33}$$

However, inference from conditionally specified models is not direct because the normalizing constant is an unknown function of the parameters. The shape of the likelihood is known but not the factor required to make it integrate to 1. A method to avoid dealing with the normalizing constant consists of using both conditional distributions. We define the pseudolikelihood estimate of $\underline{\theta}$ to be that value of $\underline{\theta}$ which maximizes the pseudolikelihood function defined by:

$$\tilde{l}(\underline{\theta})=\prod_{i=1}^n f_{X|Y}(x_i \mid y_i;\underline{\theta})f_{Y|X}(y_i \mid x_i;\underline{\theta}), \tag{34}$$

According to Arnold and Strauss [20] these estimators are consistent and asymptotically normal. In this kind of conditional models, these estimators are much easier to obtain than the maximum likelihood estimates.

6. Applications

The model with normal conditionals can present several modes and in consequence is a natural alternative to mixture models for modelling heterogeneity and also can be used for modelling a population composed for several cluster. Arnold, Castillo and Sarabia [21] used this bivariate distribution for fitting the classical Fisher data where there are pooled two different samples. The model was fitted by pseudo-likelihood.

The model with lognormal conditionals has been used by Sarabia, Castillo, Pascual and Sarabia [19] for modelling bivariate income distributions, using the information contained in the European Community Household Panel. These authors have used the Spanish microdata (approximately 10,500 individuals), focusing analysis on waves 1, 3 and 6. It is important to point out that are a big number of bivariate data with high variability. They fitted to these two sets of data the classical bivariate lognormal distribution and the bivariate lognormal conditional distribution (22) with $\delta_i = 0$, maximizing the pseudo-likelihood function given in (34). The resulting fitted model is very acceptable and implies a very significant improvement in the fit of the bivariate lognormal conditional distribution.

7. An extension: Bivariate distributions with skew-normal conditionals

Several extension of the previous models are possible. Bivariate and multivariate distributions with t-Student conditionals were studied by Sarabia [22]. Sarabia, Castillo, Pascual and Sarabia [19] proposed several extensions of the bivariate distribution with Lognormal conditionals given by (21). In this section we review models with Skew-Normal conditionals that were studied by Arnold, Castillo and Sarabia [23]. The univariate skew-normal distribution is a class of distributions whose density takes the form

$$f(x;\lambda) = 2\,\phi(x)\,\Phi(\lambda x), \quad x \in \Re, \tag{35}$$

where $\phi(x)$ and $\Phi(x)$ denote, the standard normal density and distribution functions, respectively. The parameter $\lambda \in \Re$ is a parameter which governs the skewness of the distribution. We will write $X \sim SN(\lambda)$. The skewness of this distribution is a bit limited. In order to increase coverage of the (β_1, β_2) plane it is convenient to introduce an extra parameter and define densities:

$$f(x; \lambda, \xi) = \frac{\phi(x)\, \Phi(\lambda x + \xi)}{\Phi\left(\xi(1+\lambda^2)^{-1/2}\right)}, \quad x \in \Re . \tag{36}$$

We are interested in the form of the density for a two dimensional random variable (X, Y) such that:

$$X \mid Y = y \sim SN(\lambda_1(y)) \tag{37}$$

and

$$Y \mid X = x \sim SN(\lambda_2(x)) \tag{38}$$

for some functions $\lambda_1(y)$ and $\lambda_2(x)$. If (37)-(38) are to hold, it must exist densities $f_X(x)$ and $f_Y(y)$ such that

$$f_{X.Y}(x, y) = 2\, \phi(x)\, \Phi(\lambda_1(y)x)\, f_Y(y) = 2\, \phi(y)\, \Phi(\lambda_2(x)y)\, f_X(x) . \tag{39}$$

In this functional equation, $f_X(x)$, $f_Y(y)$, $\lambda_1(y)$ and $\lambda_2(x)$ are unknown functions to be determined. It is not hard to proof that $f_Y(y) = \phi(y)$ and $f_X(x) = \phi(x)$. Then we have:

$$\Phi(\lambda_1(y)x) = \Phi(\lambda_2(x)y), \quad \forall x, y$$

and then we get the solutions $\lambda_1(y) = \lambda y$ and $\lambda_2(x) = \lambda x$ where λ is a constant. In consequence, we have two types of solutions to the previous functional equation. The first one corresponds to the independence case. In this situation we have $\lambda_1(y) = \lambda_1$, $\lambda_2(x) = \lambda_2$, $X \sim SN(\lambda_2)$, $Y \sim SN(\lambda_1)$ and

$$f_{X.Y}(x, y) = 4\, \phi(x)\, \phi(y)\, \Phi(\lambda_2 x)\, \Phi(\lambda_1 y)$$

The second situation corresponds to the dependent case. Here $\lambda_1(y) = \lambda y$ and $\lambda_2(x) = \lambda x$ and consequently $f_X(x) = \phi(x)$, $f_Y(y) = \phi(y)$ and

$$f_{X.Y}(x, y) = 2\, \phi(x)\, \phi(y)\, \Phi(\lambda xy). \tag{40}$$

The previous joint density has standard normal marginals together with skewed normal conditionals. The corresponding regression functions are non-linear and take the form:

$$E(X \mid Y = y) = \sqrt{\frac{2}{\pi}} \cdot \frac{\lambda y}{\sqrt{1 + \lambda^2 y^2}} .$$

The correlation coefficient is:

$$\rho(X,Y) = \text{sign}(\lambda) \times \frac{U(3/2, 2, 1/2\lambda^2)}{2\lambda^2\sqrt{\pi}}$$

where $U(a,b,z)$ represents the Confluent Hypergeometric function. It can be shown that $|\rho(X,Y)| \leq 0.63662$. Again, multimodality is possible. If $|\lambda| \leq \sqrt{\pi/2} \approx 1.25$, the density (40) has a unique mode at the origin, $(0,0)$, and if $|\lambda| > \sqrt{\pi/2}$, the density (40) is bimodal. More complicated models based on the density (36) have been considered by Arnold, Castillo and Sarabia [23].

References

1. B.C. Arnold and S.J. Press. (1989). Compatible conditional distributions. Journal of the American Statistical Association, 84, 152–156.
2. E. Castillo and J. Galambos. (1987). Bivariate distributions with normal conditionals. Proceedings of the International Association of Science and Technology for Development, 59–62. Anaheim, CA: Acta Press.
3. E. Castillo and J. Galambos. (1989). Conditional distributions and the Bivariate normal distribution. Metrika, 36, 209–214.
4. A. Bhattacharyya. (1943). On some sets of sufficient conditions leading to the normal Bivariate distribution. Sankhya, 6, 399–406.
5. B.C. Arnold, E. Castillo and J.M. Sarabia. (1999). Conditional specification of statistical models. Springer Series in Statistics. New York: Springer Verlag.
6. M. Ahsanullah. (1985). Some characterizations of the Bivariate normal distribution. Metrika, 32, 215–218.
7. B.C. Arnold, E. Castillo and J.M. Sarabia. (1994a). A conditional characterization of the multivariate normal distribution. Statistics and Probability Letters, 19, 313–315.
8. B.C. Arnold, E. Castillo and J.M. Sarabia. (1994b). Multivariate normality via conditional specification. Statistics and Probability Letters, 20, 353–354.
9. W. Bischoff. (1993). On the greatest class of conjugate priors and sensitivity of multivariate normal posterior distributions. Journal of Multivariate Analysis, 44, 69–81.
10. W. Bischoff. (1996a). Characterizing Multivariate Normal Distributions by Some of its Conditionals. Statistics and Probability Letters, 26, 105–111.
11. W. Bischoff. (1996b). On distributions whose conditional istributions are normal. A vector space approach. Mathematical Methods of Statistics, 5, 443–463.
12. W. Bischoff and W. Fieger. (1991). Characterization of the multivariate normal distribution by conditional normal distributions. Metrika, 38, 239–248.

13. T.T. Nguyen, G. Rempala and J. Wesolowski. (1996). Non-Gaussian measures with Gaussian structure. Probability and Mathematical Statistics, 16, 287–298.
14. A. Gelman and X.L. Meng. (1991). A note on Bivariate distributions that are conditionally normal. The American Statistician, 45, 125–126.
15. J.M. Sarabia. (1995). The centered normal conditionals distribution. Communications in Statistics, Theory and Methods, 24, 2889–2900.
16. B.C. Arnold, E. Castillo, J.M. Sarabia and L. Gonzalez-Vega. (2000). Multiple modes in densities with normal conditionals. Statistics and Probability Letters, 49, 355–363.
17. P.W. Holland and Y.L. Wang. (1987). Dependence function for continuous Bivariate densities. Communications in Statistics, Theory and Methods, 16, 863–876.
18. M.C. Jones. (1996). The local dependence function. Biometrika, 83, 899–904.
19. J.M. Sarabia, E. Castillo, M. Pascual and M. Sarabia. (2005). Bivariate income distributions with lognormal conditionals. International Conference in Memory of Two Eminent Social Scientists: C. Gini and M.O. Lorenz, 23–26.
20. B.C. Arnold and D. Strauss. (1988). Pseudolikelihood estimation. Sankhya, Ser. B, 53, 233–243.
21. B.C. Arnold, E. Castillo and J.M. Sarabia. (2001). Conditionally specified distributions: An introduction (with discussion). Statistical Science, 16, 151–169.
22. J.M. Sarabia. (1994). Distribuciones Multivariantes con Distribuciones Condicionadas t de Student. Estadística Española, 36, 389–402.
23. B.C. Arnold, E. Castillo and J.M. Sarabia. (2002). Conditionally specified multivariate skewed distributions. Sankhya, A, 64, 1–21.

Chapter 11

INEQUALITY MEASURES, LORENZ CURVES AND GENERATING FUNCTIONS

J.J. NÚÑEZ-VELÁZQUEZ*

Departamento de Estadística, Estructura Económica y O.E I., University of Alcalá
Plaza de la Victoria, 2, 28802 Alcalá de Henares (Madrid), Spain

This paper studies the foundations of income inequality measures and its relations with Lorenz curves, the Pigou-Dalton transfer principle and majorization relations among income vectors. So, the historic development of these concepts is surveyed to see how the actual set of properties and *axioms* was generated, in order to define when an inequality measure has a good perform. Finally, this work includes an analysis studying the problem associated with inequality orders and dominance relations among income vectors.

1. Introduction

It may be considered that the interest raised in the last thirty years in the researcher's community, related to the study of economic inequality aspects has begun since the seminal paper by Atkinson (1970) and the book by Sen (1973) as its main focuses. Both of them have had profound effects on this research field. Since then, papers and books on this task appear frequently in the economic literature and this root interest has been spread to several nearby important social problems, like poverty, mobility, polarization and privation studies, among others.

In this period of time, different approximations to this problem have been developed, including social welfare assumptions from Economic Theory to support several economic inequality measures[a]. However, the number and variety of these assumptions have considerably increased in such a way that some of them have been matter of hard controversy. Some outstanding examples

---

*This work is dedicated to the memory of Camilo Dagum, recently died. He was a direct disciple of C. Gini and a master of several generations of researchers.

[a]When inequality measures are referred, we must understand them as functions or indicators defined over an income distribution. So, these indicators are supposed to *measure* how much inequality is present in the resources sharing. In other words, there are no connections with the same commonly used concept in *Measure Theory*. So, along the paper, we shall use the words indicator and measure in an interchangeable manner.

of these works could be Cowell (1995), Foster (1985), Nygard and Sandström (1981) or Dagum (2001), among others. In the Spanish case, we would quote the works published by Zubiri (1985), Ruiz-Castillo (1987) or Pena et al. (1996).

Nevertheless, despite the huge amount of related literature, Lorenz curve paradigm remains nowadays as the cornerstone of economic inequality analysis. Indeed, Lorenz curve should be considered as the basic tool to be taken into account to support inequality analysis, even though this proposal was presented by Lorenz (1905), more than a century ago. Along all this time, Lorenz curve has resisted all the alternative proposals suggested to modify it.

Because of the above argument, one of the main objectives of this paper must be to pay tribute to Lorenz, a century after his curve's proposal. To put Lorenz curves in context, a description of the 9 pages long original paper is quoted from Arnold (2005), which was pronounced at the Siena Congress, just celebrated owing to the commemoration of such an event. He wrote: ... *In the last 3 pages of the paper he describes what will become the Lorenz curve. Actually there are only 35 lines of text and two diagrams devoted to the topic. It has all grown from that!* ...

First of all, in this paper, we review the classical concepts related to income majorization, in order to identify the theoretical background underlying Lorenz curves and economic inequality measures in the way we understand them nowadays. This aim should be justified because we must reconsider what the underlying basic concepts are really imbedded under economic inequality measurement. In doing so, it would result in a better comprehension about what elements are playing a significant role when economic inequality is intended to be measured. Moreover, the aforementioned understanding must allow us to back up an efficient selection about which the better inequality measures could be. In this sense, a set of properties will be proposed in order to analyze the suitability of a huge amount of economic inequality measures. Additionally, a brief analysis of other related concepts and methods, recently proposed, will be included. Variety of themes this paper deals with, advice us to provide the paper with a well-disaggregated structure, which is exposed next.

So, the paper is structured as follows. In section 2, a brief chronology of published concepts related to economic inequality is developed, emphasizing those which are close to Lorenz curves methodology. Section 3 is devoted to set the basic framework with respect to income distribution space and to present the crucial majorization concepts. Section 4 studies, on the one hand, the meaning of economic inequality and, on the other hand, it is dedicated to Lorenz curves methodology to analyze income inequality as well as connected methods like direct functional forms estimation or the less-known generating functions. Section 5 presents the Pigou-Dalton Transfer Principle, as a key element to play a role in economic inequality; and enlightens its relation with majorization comparisons. Section 6 connects income inequality measurement with Schur-convex or S-convex functions, closing the circle of relationships among analytical concepts exposed before. However, section 7 intends to face another point of view in economic inequality analysis, as axiomatic approach may be considered; nevertheless, it will be shown how, in essence, it constitutes another way of expressing the same ideas. Section 8 carries out the discussion about alternative inequality comparison criteria suggested in the literature. Once all these elements has been discussed, a group of properties is proposed to be used in selecting economic inequality measures in section 9, and it will be proved how some well-known indicators fail to fulfil them. Finally, the paper ends summing up the outstanding conclusions.

2. Brief historical evolution of concepts related to economic inequality analysis

First studies about economic inequality of income distributions must to be related to the majorization relationship between a pair of them. So, Muirhead (1903) establishes a relation of majorization concept to progressive income transfers, which will be expressed in formal terms later. In the year 1905, M.O. Lorenz proposes his curves to analyze income and wealth inequality and he points out that his curve's bow is an indicator of the inequality degree included in the distribution.

In 1912, C. Gini proposes his indicator to measure inequality, using the mean difference measure obtained by averaging the differences between every pair of incomes from the distribution. In the same year, A.C. Pigou suggests the ideas which lately will be stated as the Transfer Principle, when H. Dalton expressed them into rigorous terms in the year 1920. This principle is one of four well-known properties, including the so-called Dalton Population Principle.

Related to the majorization concept again, I. Schur proposed his convexity concept in the year 1923. This concept is strongly closed to bi-stochastic matrices and therefore near to the progressive transfer concept too.

In 1929, G.H. Hardy, J.E. Littlewood and G.Polya publish their first results about inequality by means of an article included in the journal *The Messenger of Mathematics*. It can be considered as a precedent of his seminal book *Inequalities*, whose first edition appeared in the year 1934. However, in 1932, J. Karamata proves the theorem which is named after him since then, although

Hardy, Littlewood and Polya had proposed it in 1929. Its content can be regarded as one of the cornerstones of economic inequality measurement.

In the year 1979, J. Gastwirth proposes the explicit expression of general Lorenz curves, allowing the use of random variable-based ones, whatever its type would be.

Obviously, this brief review must contain a mention to the aforementioned paper by A.B. Atkinson, in 1970, where he sets key arguments on the normative content of inequality measures through a family of indicators named after him. These arguments are based on the general mean function or generalized mean, but they are not free of controversy. Again, it is necessary to make reference to the appearance of the book *On Economic Inequality,* by A.K. Sen in 1973, which has been reedited in 1997, including a wide annexe with several advances in economic inequality and poverty registered during the elapsed time of 25 years between them. This new annexe has been written by the same author with J.E. Foster.

Precisely, J.E. Foster published, in 1985, his renowned theorem, where he determined the conditions an indicator has to fulfil to be compatible with the order generated using the Lorenz curve. These conditions impose suitable properties on inequality measures in order to reach a performance according to Lorenz curves do and they are conceptually different from the aforementioned normative ones. This result constitutes the basic system of properties which are required an inequality indicator to achieve and it may be considered as a starting point in the search of relevant properties, so-called *inequality axioms,* to select an adequate indicator. Nevertheless, this way of choosing an inequality indicator had some precedents in the literature.

Finally, in 2001, C. Dagum publishes in the Spanish journal *Estudios de Economía Aplicada* a summary from several papers published before in different journals, since 1981. In this work, the author exposes his point of view about the economic foundations of different inequality measures in contrast to the normative view derived through the Atkinson's approach[b].

Along this necessarily brief revision, we have tried to point out the evolution that economic inequality study has registered, taking into account those several concepts configured as fundamentals on this subject treatment. Although nowadays these contents are usually presented as properties or *axioms* like it was explained before, we believe this paper will show the links among these properties and basic concepts supporting them. In the following sections, the aforementioned concepts will be developed.

---

[b]A more detailed description of this point of view can be seen in Dagum (1990).

3. Income distribution space and the majorization concept[c]

Firstly, we are going to define the income distribution space as a support tool for the remaining concepts. So, if the population contains N individuals, an income distribution will be any real multidimensional vector from $\mathbf{R}^N$, provided that all its components must be non negative so that there exists something to share among the individuals compounding the population. More precisely:

$$D_N^* = \left\{(x_1, x_2, \ldots, x_N) : x_i \geq 0, i = 1, \ldots, N; \sum_{i=1}^{N} x_i > 0\right\} \tag{1}$$

But attending only to the inequality in the distribution, every vector permutation provides the same resource sharing, whatever would be the identity or the place corresponding to each single individual[d]. To formalize this idea, let Π_{NxN} be the N-order permutation matrices set and let us define the following equivalence relation over D_N^*:

$$x \approx y \Leftrightarrow x = \Pi \cdot y, \; \Pi \in \Pi_{NxN} , \tag{2}$$

so that we shall choose the ordered income vector, from smallest to largest, as the canonical element of each equivalence class. Thus:

$$D_N = D_N^* / \approx \; = \{(x_1, x_2, \ldots, x_N) : x_1 \leq x_2 \leq \ldots \leq x_N\} \tag{3}$$

Then, the income distribution space to be considered will be:

$$D = \bigcup_{N=2}^{\infty} D_N$$

Now, we can go on to define the majorization relation between income distributions belonging to D. Let *x*, *y* be two income distributions belonging to D_N; then we shall say that *x* is majorized by *y* and write $x \prec y$ when it is more equally distributed. So:

---

[c]Here, we are referring only to the economic concept of income, although the analysis can be applied directly to other concepts related to the individual or household economic positions, like earnings, expenditures or wealth. However, there is controversy about what the economic position must be used, because of both theoretical grounds and disposable data reliability (Ruiz-Castillo, 1987; Pena et al., 1996, among others).

[d]This argument is usually known as *symmetry or anonymity axiom* related to inequality measurement (Foster, 1985).

$$x \prec y \Leftrightarrow \begin{cases} \sum_{i=1}^{k} x_i \geq \sum_{i=1}^{k} y_i, & k = 1,2,\ldots,(N-1) \\ \sum_{i=1}^{N} x_i = \sum_{i=1}^{N} y_i \end{cases} \tag{4}$$

It is easy to note how this relation constitutes the straight precedent of Lorenz curves comparison between pairs of income distributions, as we shall present later. However, majorization turns to be a more restrictive relationship because it only allows comparisons between income distributions defined over equally sized populations and where the total amount of shared resources has to be the same. It is trivial to prove that this relation defined over D_N presents a partial order or quasi-order structure.

4. Economic inequality, Lorenz curves and generating functions

An early precedent of the inequality concept can be found when V. Pareto (1897) identifies a *smaller inequality* with a situation where personal incomes tend to be more similar. Castagnoli and Muliere (1990) points out how this argument constitutes an early version of Transfer Principle, although it is not formalized yet.

On the other hand, majorization relationship includes the more-equally distributed concept, meaning that its components are more similar than the respective ones in the other vector we are comparing with. Using this fact, it is possible to clarify definitively the inequality concept we are trying to measure. In that respect, it is very useful the following quote from S. Kuznets to describe the concept: *When we speak about income inequality, we are simply referring to income differences, without taking into account its desirability as a reward system or its undesirability as a system contradicting a certain equality scheme.* (S. Kuznets, 1953, pg. xxvii). According to the previous statement, an economic inequality measure is not supposed to judge if the sharing is adequate or not but to quantify if the actual distribution is near or far from the equality situation, where all components in the population perceives the same income, although this fact has not to be a goal itself. In such a sense, Bartels (1977) demands a reference distribution in order to an inequality measure could compare with. Although such a reference might be what society considers a fair distribution, this approach did not find support in the literature because of the inherent difficulty involved in the reference distribution determination. Therefore, the usual reference distribution to compare with turns to be the egalitarian one, where all its components are the same and equal to mean income.

And still inequality assertions generate great repercussions, as A.K. Sen points out: *The idea of inequality is both very simple and very complex. At one level it is the simplest of all ideas and has moved people with an immediate appeal hardly matched by any other concept. At another level, however, it is an exceedingly complex notion which makes statements on inequality highly problematic, and it has been, therefore, the subject of much research by philosophers, statisticians, political theorists, sociologists and economists* (A.K. Sen, 1973, pg. vii). Of course, thirty three years later, it can not be said that exists an agreement among researchers about a consensus method to measure inequality, while the nearest instrument to this situation comes to be the Lorenz curve we are going to present now.

4.1. *Lorenz curves and the Lorenz dominance criterion*

The curve, proposed by Lorenz (1905), can be defined in the following way. Let x be an income distribution from D. Using its ordered components, cumulative relative frequencies of individuals and resource shares are calculated, keeping in mind that they are non negative. Also, let $\overline{x}$ be the income mean; then:

$$\left.\begin{aligned} p_0 = 0;\ p_i = \frac{i}{N},\ i = 1,2,\ldots,N \\ q_0 = 0;\ q_i = \frac{1}{N\overline{x}}\sum_{j=1}^{i} x_j\ ,i = 1,2,\ldots,N \end{aligned}\right\} \qquad (5)$$

Thus, the Lorenz curve, L(p), is obtained by linking the points contained in the set $\{(p_i,q_i);\ i = 0,1,\ldots,N\}$, using linear interpolation to generate a polygonal curve. Obviously, L(p) is inscribed within the unit square. So, if L(p) is near to the unit square's diagonal, then the income sharing will be near to the egalitarian situation. Else, the more bent the bow's curve is, the more inequality will be present in the income distribution.

The previous definition is a descriptive one, but it can be easily generalized to the case when we are dealing with a non-negative random variable, X, to model incomes. In such a case, let μ be its expectation E(X) and let F(x) be its cumulative distribution function. Now, definition (5) can be expressed as (Kendall and Stuart, 1977, for example):

$$\left.\begin{aligned} p = F(x) = \int_0^x dF(t) \\ q = L[F(x)] = \frac{1}{\mu}\int_0^x t\cdot dF(t) \end{aligned}\right\} \qquad (6)$$

In this context, Gastwirth (1971) suggests an integrated framework,, allowing us to express Lorenz curve into an explicit general way:

$$L(p) = \frac{1}{\mu} \int_0^p F^{-1}(t) \cdot dt \qquad (7)$$

where $F^{-1}(p) = \inf\{x : F(x) \geq p\}$.

Lorenz curve properties are very well-known[e], but it is worth standing out that if L(p) is derivable, its slope will be given by:

$$t(p) = \frac{F^{-1}(p)}{\mu}, \quad p \in (0,1) \qquad (8)$$

Also, the difference (related to the diagonal) function will be:

$$A(p) = p - L(p), \quad p \in [0,1] \qquad (9)$$

and it reaches a maximum at the point p = F(μ). Moreover, it is particularly interesting the following result[f]:

Theorem 1 (Iritani and Kuga, 1983): Let q = L(p) be a function defined over the interval [0,1]. Then, L(p) is a Lorenz curve corresponding to some non-negative random variable X, if and only if L(p) satisfies the following properties:

$$L(0) = 0, L(1) = 1.$$

L(p) is convex and non-decreasing.

However, the precedent discussion about Lorenz curves suitability had as a main objective making inequality comparisons between income distributions. To accomplish this aim, the following relationship, called *Lorenz dominance criterion* is going to be established.

Definition 1: Let $x, y \in D$. Then x is said to be *less unequal than y in the Lorenz sense* ($x \leq_L y$) when the Lorenz curve associated to y contains completely the corresponding to x. Formally:

$$x \leq_L y \Leftrightarrow L_x(p) \geq L_y(p), \quad \forall p \in [0,1] \qquad (10)$$

Related to majorization relation, Lorenz criterion turns out to be a more versatile relation because of its capability of making comparisons between

[e]See, for example, Casas and Núñez (1987) or Nygard and Sandström (1981), for more details.

[f]Analyses of sampling results about Lorenz curves are out of the scope of this paper. However, there are very interesting references in this field, beginning with Goldie (1977) on strong consistency of empirical Lorenz curves, and Beach and Davidson (1983) or Beach and Richmond (1985) on asymptotic normality of Lorenz curves estimates.

income vectors coming from different-sized populations. Nevertheless, in this last case, it becomes evident that:

$$x \leq_L y \Leftrightarrow \left(\frac{x}{\bar{x}}\right) \prec \left(\frac{y}{\bar{y}}\right); \quad x, y \in D_N \tag{11}$$

Thus, what it is under the Lorenz dominance criterion is just the majorization concept. Now, Lorenz relationship presents a pre-order structure (reflexive and transitive properties), whenever it be defined between proportional income distribution classes (or partial order if not). Therefore, if two Lorenz curves cross each other, the associated income distributions will be no comparables and this is a very frequent situation in practice, giving an incomplete inequality ranking as a result. In applications, this structure used to be plotted using the so-called *Hesse diagrams,* as it can be seen in Pena et al. (1996), for example.

Actually, the absence of a total order is the main reason to use inequality measures, in order to overcome the lack of ranking in an appreciable number of paired comparisons and to quantify inner inequality levels too. Inequality indicators induce a total order because its values are real numbers but the price we have to pay is the inclusion of underlying weighting schemes on the distributions, and these are not always clear enough to deduce. So, several inequality indicators may produce different rankings on the implied distributions.

4.2. *Parametric estimation of Lorenz curves*

Theorem 1 turns to be a very important result because of its characterization of Lorenz curves. Moreover, the researcher may decide sometimes to estimate a parametric Lorenz curve directly from data (p_i,q_i). In such a case, it is strongly necessary to know what the possible parametric functional forms could perform like a Lorenz curve and Theorem 1 shows what the required properties must be. Recently, this adjusting procedure has constituted an active researching field, whose guidelines are presented below.

Some of the simplest functional forms used as a parametric Lorenz curve are the following ones:

i) Potential: $L(p) = p^b$, $b \geq 1$ (e.g., Casas and Núñez, 1991)
ii) Exponential: $L(p) = p.a^{(p-1)}$, $a > 1$ (e.g., Gupta, 1984)
iii) Potential-Exponential: $L(p) = p^b.e^{-c(1-p)}$, $b \geq 1$, $c > 0$ (Kakwani and Podder, 1973).

There exists a great deal of more complex functional forms. Furthermore, it can be proved the next assertion: If some Lorenz curves satisfy the conditions included in Theorem 1, then every convex mixture of them will fulfil such conditions too. (Casas, Herrerías and Núñez, 1997). In other words, every convex mixture of Lorenz curves turns out to be another Lorenz curve. So, we have an infinite number of possible functional forms for estimating Lorenz curves.

In addition, there is another method to generate new functional forms capable to estimate Lorenz curves indeed. Now, the procedure consists in obtaining new functional Lorenz curves by applying specific transformations over an original one.

Theorem 2 (Sarabia, Castillo and Slotjje, 1999): Let L(p) be a Lorenz curve. Then, the next transformations generate Lorenz curves too[g]:

a) $L_\alpha(p) = p^\alpha . L(p)\ ,\ \alpha \geq 1.$

b) $L_\alpha(p) = p^\alpha . L(p)\ ,\ 0 \leq \alpha < 1, L''_\alpha(p) \geq 0.$

c) $L_\gamma(p) = L(p)^\gamma\ ,\ \gamma \geq 1.$

Moreover, another approach in this field consists of imposing directly parametric cumulative distribution functions over the income. In this respect, examples of such models used in practice are the Wakeby distribution (Houghton, 1978), generalized Tukey lambda (Ramberg, et al., 1979) or Mc Donald (Sarabia, Castillo and Slottje, 2002).

4.3. *Generating functions and Lorenz curve functional forms*

In this section, we start with the definition of the density generating function, to go on with its generalization related to Lorenz curves. Finally, we are going to explore the relationship both types of generating functions.

So, the (density) generating function associated to every continuous random variable with f(.) as its probability density function may be defined as (Callejón, 1995):

$$g_F(x) = \frac{f'(x)}{f(x)} \tag{12}$$

This generating function allows us to obtain ordered families of Lorenz curves. Such ordered families only depend on a parameter and, therefore, give us a total order structure on paired comparisons using Lorenz dominance

[g]Regarding estimation metjods in such a matter, see e.g. Castillo, Hadi and Sarabia (1998).

criterion, and this fact is due to the only parameter they have. The simplest case corresponds to *strongly unimodal distributions* or, in other words, those whose probability density function is log-concave. That is:

$$\frac{d}{dx}\left(\frac{f'(x)}{f(x)}\right) = g'_F(x) \leq 0, \quad \forall x \in R \tag{13}$$

If the support of the random variable X consists on an interval (a,+∞), Lorenz curves derived from this definition will be (Arnold et al., 1987):

$$L_\tau(p) = F(F^{-1}(p) - \tau)\,,\ \tau \geq 0 \tag{14}$$

Some examples of this kind of random variables are log-normal and Pareto distributions. As it may be seen through the mentioned examples, the main drawback with these distributions is the rigidity as real income models[h].

On the other hand, in the same way as before, the Lorenz curve generating function can be defined *mutans mutandi,* assuming now L(.) as a Lorenz curve from a continuous random variable:

$$g_L(p) = \frac{L'(p)}{L(p)} \Rightarrow L(p) = k.\exp\left(\int g_L(p).dp\right) = k.\,e^{G(p)}\ , \tag{15}$$

but, in this case, the obtained function might not be a Lorenz curve because of its additional properties. So, we need to establish what restrictions must be imposed on that definition. To answer this question, it can be easily proved the next result:

Proposition 1 (Herrerías, Palacios and Callejón, 2001): In the same above circumstances, L(p) is a Lorenz curve if and only if the following conditions are satisfied:

a) $k = \exp(-G(1))$

b) $\lim_{p \to 0^+} G(p) = -\infty$

c) $g_L(p) > 0, \ \forall p \in (0,1]$

d) $(g_L(p))^2 + g'_L(p) > 0, \ \forall p \in (0,1]$

So, if a function $g_L(p)$ fulfils the above conditions, then it will give a Lorenz curve through the associated generating function. Using several generating functions, García and Herrerías (2001) has obtained a number of well-known

---

[h]Although more complex, another method to generate ordered families of Lorenz curves can be seen in Sarabia, Castillo and Slotjje (1999).

functional forms, corresponding to Lorenz curves associated to some probabilistic income models.

However, the relationship between Lorenz curves and probability density function is so close than we would expect a readily explicit relationship between both generating functions, but this is not the case. The aforementioned relationship turns out to be hard to accomplish. So, if implied functions are sufficiently differentiable, then we can only prove the next equations system, using the same notation as above:

$$\left.\begin{aligned} g_F(x) &= E(X).f'(x).L''[F(x)] \\ g_L[F(x)] &= \frac{x}{E(X).L[F(x)]} \end{aligned}\right\} \qquad (16)$$

5. Majorization and the Pigou-Dalton transfer principle

Despite the informal precedent appeared in Pareto (1987), it was H. Dalton (1920, pg. 351) who stated the Transfer Principle, from the guidelines exposed by Pigou (1912, pg. 24): *If only there are two income receptors, and a transfer from the richest to the poorest is produced, inequality must decrease.* Below, he imposes the transferred account must not change the relative positions of both involved individuals as an obvious restriction and he concludes that the most equalizer transfer is half of the income difference between them.

In a general version, the Pigou-Dalton Transfer Principle can be established as follows: *If an income distribution y is obtained from x by a progressive (regressive) transfer, or a non-empty sequence of them, then inequality decreases (increases).* Now, we can state the progressive transfer concept, from a more rigorous point of view.

Definition 2: Let $x, y \in D_N$. Then, y is said to be obtained from x through a progressive income transfer if:

$$x = (x_1,\dots,x_i,\dots,x_j,\dots,x_N)' \Rightarrow y = (x_1,\dots,x_i+\delta,\dots,x_j-\delta,\dots,x_N)', \quad \delta \in \left(0, \frac{x_j - x_i}{2}\right)$$

In such a case, x is said to be obtained from y through a regressive transfer.

Next, the link between this just introduced concept and majorization relationship is going to be analyzed. A pioneer result published in 1903 serves this purpose and it can be stated by means of the presented terminology as follows:

Theorem 3 (Muirhead, 1903): Let $x, y \in D_N$. Then, x is majorized by y ($x \prec y$) if and only if x can be obtained from y through a finite number of progressive transfers.

Therefore, we must conclude that the Pigou-Dalton Transfer Principle represents the essence of majorization relationship defined over pairs of income distributions, and then also of dominance criterion in the sense of Lorenz and, furthermore, of economic inequality. However, in spite of the importance of the previous assertion, we must admit this formulation as scarcely operative, to a certain extent. So, the following objective will be to achieve an effective characterization of both concepts.

To fulfil this aim, we appeal to bi-stochastic matrices set, which definition is presented below.

Definition 3: A matrix P_{NxN} is said to be *bi-stochastic* or *doubly stochastic* if it satisfies the following properties:

$$\begin{aligned} &\text{i) } 0 \leq p_{ij} \leq 1, \quad \forall i,j = 1,2,\ldots,N \\ &\text{ii) } \sum_{j=1}^{N} p_{ij} = 1, \quad \forall i = 1,2,\ldots,N \\ &\text{iii) } \sum_{i=1}^{N} p_{ij} = 1, \quad \forall j = 1,2,\ldots,N \end{aligned} \tag{17}$$

Thus, doubly stochastic matrices are finite ones with a probability distribution defined over each row or column. The set including all these matrices is closely related to permutation matrices, in the way expressed by the following result.

Theorem 4 (Birkhoff, 1976): The (NxN) bi-stochastic matrices set constitutes the convex envelope of the (NxN) permutation matrices set.

Furthermore, it would be easy to prove how the application of a doubly stochastic matrix over an income distribution produces an equalizing effect. It is enough to let P be a bi-stochastic matrix and $x, y \in D_N$, so that $x = P.y$; then each component of vector x will be a convex mixture of the vector y components and thus we have a progressive transfer[i]. In other words:

$$x_i = \sum_{j \neq i} y_j p_{ij} + y_i \left(1 - \sum_{j \neq i} p_{ij} \right) = y_i + \sum_{j \neq i} \left(y_j - y_i \right) p_{ij}, \quad i = 1,2,\ldots,N \tag{18}$$

---

[i] In that sense, Arnold (2005), quoting from Schur (1923), refers them defining x as an averaging of y.

The anterior explanation is sufficient to make evident the following result, despite the great advance it has represented in this field.

Theorem 5 (Hardy, Littlewood and Polya, 1959):

$$(x \prec y) \Leftrightarrow x = P.y, \quad \forall x, y \in D_N,$$

where P is some (NxN) doubly stochastic matrix.

Taking into account the results exposed before, progressive income transfers have been characterized through operations involving income vectors and bi-stochastic matrices. So, progressive transfers have been reduced to algebraic operations, making its treatment easier.

6. Inequality measures and Schur convexity

The main objective of this section is to obtain functions compatible with the majorization relation, in order to construct inequality measures over income distributions. In doing so, it must be taken into account the links with Transfer Principle and Lorenz dominance. These functions are defined below.

Definition 4: A real function $\varphi(.)$, defined over DN is called Schur convex or S-convex when it is monotone with respect to majorization relationship. Formally:

$$(x \prec y) \Rightarrow \varphi(x) \leq \varphi(y) \tag{19}$$

If there were strict inequality, the function would be called *strictly S-convex.*

A useful characterization of such a function is the one contained in the next result, which makes easier its manipulation.

Theorem 6 (Schur and Ostrowski, 1952): Let I be a real interval and $\varphi(.)$ a continuously differentiable function defined over I^N. Then, $\varphi(.)$ will be S-convex if and only if it satisfies the following conditions:

i) $\varphi(.)$ is symmetric over I^N.

ii) Schur condition:

$$(x_i - x_j)\cdot\left[\frac{\partial\varphi}{\partial x_i}(x) - \frac{\partial\varphi}{\partial x_j}(x)\right] \geq 0, \quad \forall i \neq j, \ \forall x \in D_N \cap I^N \tag{20}$$

Furthermore, it can be proved that every convex and symmetric function is S-convex too (Marshall and Olkin, 1979, pg. 67).

From now on, it becomes evident how inequality measures should be S-convex functions defined over income distributions, keeping in mind all the equivalences stated before. For example, Gini index (Gini, 1912) is a strictly S-convex function[j]. However, usual inequality measures construction is based on the next statement, which connects all the implications related to inequality and majorization exposed before.

Theorem 7 (Karamata, 1932): Let g(.) be a convex, continuous and real function, then:

$$(x \prec y) \Leftrightarrow \sum_{i=1}^{N} g(x_i) \le \sum_{i=1}^{N} g(y_i), \quad \forall x, y \in D_N$$

Further, if g(.) is a convex real function, then $h(x) = \sum_{i=1}^{N} g(x_i)$ is said to be a *convex separable* function, provided that $x \in D_N$.

It is easy to see that every convex separable function is S-convex too. Nevertheless, the inverse statement is not true and this can be readily checked from Theorem 7. But it is important to observe how Theorem 7 relates majorization to economic inequality measures construction. Moreover, the following property links this to Lorenz dominance.

Corollary 1 (Arnold, 1987): Let g(.) be a convex, continuous and real function, then:

$$x \le_L y \Leftrightarrow E\left[g\left(\frac{x}{E(x)}\right)\right] \le E\left[g\left(\frac{y}{E(y)}\right)\right]$$

As a result, it is worth mentioning that each convex, continuous and real function can generate a genuine inequality indicator, because it will be compatible with Lorenz dominance criterion, using Corollary 1. So, the partial order deduced from Lorenz dominance criterion is still present, connecting to the *intersection quasi-order* (Sen, 1973, pg. 72), which constitutes another partial order rather less restrictive[k].

Evidently, choosing a single inequality measure implies a total order as a result, but Lorenz compatibility hides what the causes of different orders may be when several inequality indicators are used. Reasons explaining this fact must be explained by distinct weighting schemes placed on income distribution, which are associated to each inequality measure. So, a research field has emerged, considering batteries of inequality indicators instead of choosing only one of

---

[j]Marshall and Olkin (1979) contains an extensive exposition about S-convex functions, including the result covered in Theorem 6.

[k]Obviously, this will be true only if all of the considered inequality indicators are compatible with Lorenz relation. In other case, there is no inclusion relationship linking both partial orders.

them, in order to extract the common information included in such a set using Principal Component Analysis or to eliminate the redundant inequality information through Ivanovic-Pena DP2 distance (García et al., 2002). This new approach can be modified to allow dynamic inequality evaluations too (Domínguez and Núñez, 2005).

On the other hand, Corollary 1 allows comparisons between income distributions from different-sized populations. However, this achievement is possible using homogeneous functions as inequality measures, so as proportional income vectors must give the same value. This formal fact is equivalent to impose the so-called *Dalton Population Principle*, proposed by the aforementioned author with the name Individuals Proportional Addition Principle (Dalton, 1920, pg. 357): *Inequality becomes invariant against population replicas*[1]. In formal terms, this restriction imposes that inequality measures have to be functions defined over the empirical accumulative distribution function.

Finally, we can summarize a great part of the last discussion by reproducing the next statement, where relative sensibility to income transfers is included, depending on the chosen inequality measure.

Theorem 8 (Atkinson, 1970; Kakwani, 1980): If V(.) is a strictly convex and real function, then every inequality measure defined by $I(x) = E[V(x)]$ will satisfy the Pigou-Dalton Transfer Principle, whatever the income level be. Furthermore, if V(.) is differentiable too, then its relative sensibility to income transfers will be proportional to:

$$T(x) = V'(x) - V'(x - \delta)\,, \quad \delta > 0.$$

7. Axiomatic approach to economic inequality

This approach consists of stating desirable properties a good inequality measure should fulfil, in order to be chosen among all possible ones. So, the name *axiom* must be understood in such a context, but not in the mathematical sense of unchanging true. In this way, we can impose more and more restrictive properties to limit us to choose from a lesser alternatives set. The best option would be the formulation of a group of properties able enough of characterizing a single inequality measure, because allows us its selection whenever we agree with its properties. Nevertheless, this is a difficult goal to achieve.

---

[1]A r-order population replica consists of considering an income vector which repeats r times each component of the original income distribution, giving $(x_1,\overset{r)}{\ldots},x_1,x_2,\overset{r)}{\ldots},x_2,\ldots,x_N,\overset{r)}{\ldots},x_N)'$ as a result.

Along this section, we don't intend offering an exhaustive exposition of properties but only to present the most commonly accepted ones[m]. Indeed, we will include some controversial properties to clearly establish links between this approximation through axioms imposing and the analytical treatment exposed before.

In precedent sections, we have shown how the Pigou-Dalton Transfer Principle plays a fundamental role in inequality measurement. So, we are not going to repeat its statement again, though it has to be understood as included in the basic properties set[n]. Consequently, we present below the aforementioned basic properties or axioms, where I(.) stands for a real function defined over D as a basic formulation of an inequality measure.

1. Symmetry or Anonymity Axiom

Let $x = (x_1,x_2,...,x_N)' \in D$ and $y = (x_{\sigma(1)},x_{\sigma(2)},...,x_{\sigma(N)})'$, where $\sigma(.)$ is a permutation function over the set $\{1,2,...,N\}$. Then, I(x)=I(y).

2. Homothetic or Scale Invariance Axiom

$$I(\lambda x) = I(x), \quad \forall x \in D, \forall \lambda > 0$$

3. Dalton Population Principle

Let $x,y \in D$, in such a way that y is a r-order replica of x (in other words, $y = (x',x',\overset{m)}{\dots},x')'$), and its components are increasingly ordered. Then I(x) = I(y).

4. Normalization Axiom

Its weak version expresses that $I(x) \geq 0$, $\forall x \in D$. Furthermore:

$$I(x) = 0 \Leftrightarrow \exists c \geq 0 : x = (c,c,...,c)'$$

There exists a strong version of the axiom, so-called *Range normalization axiom*, where the inequality measure must to be 1 in the case of maximum inequality.

5. Constant Addition Axiom[o]

Let $x,y \in D$, so that $x = (x_1,x_2,...,x_N)'$, $y = (x_1+c, x_2+c, ..., x_N+c)'$. Then:

$$I(y) \leq I(x), \quad \forall c \geq 0.$$

---

[m]A wide exposition of proposed axioms in the economic literature can be find in Nygard and Sandström (1981) or Ruiz-Castillo (1986), for example.

[n]Axioms presented in this section are considered the basics related to Lorenz dominance. Among the omitted ones, we must mention the *additive decomposability axiom* (Bourguignon, 1979), which stands out because of its repercussion and controversy. Moreover, this axiom allows us to characterize a family of inequality measures.

[o]This axiom appears in Dalton (1920, pg. 357).

On the other hand, to connect this approximation with the more analytical developed before, we introduce the next definition.

Definition 5: A real function I(.) defined over D is said to be a *Lorenz-compatible inequality measure* when it is monotone with respect to Lorenz dominance criterion. More formally:

$$I(x) \geq I(y) \Leftrightarrow x \geq_L y \Leftrightarrow L_x(p) \leq L_y(p), \forall p \in [0,1]$$

To characterize this kind of inequality measures, we need to formalize the restrictions that were included in the analytical framework related to Lorenz dominance and this necessity leads us to the first three axioms recently exposed. The next result summarizes this reasoning.

Theorem 9 (Foster, 1985): Let I(.) be a real function defined over D. Then, I(.) is a Lorenz-compatible inequality measure if and only if it satisfies the following axioms:

i) Symmetry.
ii) Scale Invariance.
iii) Dalton Population Principle.
iv) Pigou-Dalton Transfer Principle.

As it can be readily seen, Theorem 9 is a reformulation of the whole precedent analytical exposition into axiomatic terms. However, as it may be expected, there exist a lot of Lorenz-compatible inequality measures. Among them, we have the coefficient of variation, Gini index, Atkinson's and Theil's families of measures, as a few examples.

This axiomatic approach has allowed us to express desirable properties in order to narrow the set of alternative inequality measures to choose from. Among all of them, the Pigou-Dalton Transfer Principle plays a crucial role in inequality measurement related to Lorenz dominance criterion and majorization relationship, whereas the rest of exposed axioms have a more instrumental character[p].

In addition, the restrictions these axioms impose on inequality measures might clarify some details about the underlying weighting scheme included in each indicator. So, it is generally accepted that inequality measures would tend to weight more heavily incomes near the bottom of the distribution, while the

[p]It has been suggested the use of absolute measures, instead of relative ones. This approach implies the suppression of the Scale Invariance Axiom (Moyes, 1987). Nevertheless, this kind of measures are closer to the so-called Lorenz generalized dominance relation (Shorrocks, 1983).

limit case would be configured using only the poorer income (Rawls, 1972). Therefore, this research field intends to restrict the Pigou-Dalton Transfer Principle by placing more weighting on transfers where the smaller incomes are involved. Some related results are Shorrocks and Foster (1987) or Fleurbaey and Michel (2001), among others.

In this way of thinking, another related research field has as an objective the use of weighting schemes on the Lorenz curve directly. It is a well-known fact that Gini index matches twice Lorenz area[q] (e.g. Wold, 1935 or Kakwani, 1980). Following this idea, some authors have proposed inequality measures based on geometrical elements on the Lorenz curve, like the maximum distance to the egalitarian line (Pietra, 1914–15, 1948; Schutz, 1951), its length (Kakwani, 1980) and weighting Lorenz areas using specific functions (Mehran, 1976; Casas and Núñez, 1991, among others).

8. Alternative comparison criteria

Both majorization and Lorenz dominance relationships generate a partial order structure over D_N and D, respectively. Obviously, this fact constitutes an important drawback because it is well-known that Lorenz curves crosses occur frequently in practice (Shorrocks, 1983) and, therefore, the number of non comparable pairs of income distributions may be relatively great. So, other alternative comparison criteria have been proposed looking for a lesser number of non comparable situations, admitting that inequality is, in essence, a quasi-order (Sen, 1973) and the only way to achieve a total order is by using inequality measures, as it was shown previously. Really, it can be thought that the partial order as a result must be inherent to the problem of order relations in vector spaces.

Along this section, we will expose some of the most studied proposals in this researching field. To begin with, Shorrocks (1983) proposed the use of *generalized Lorenz curves,* claiming that they reduce significantly the number of non comparable pairs of income vectors with respect to Lorenz dominance criterion. To reach this aim, he defined his generalized curves re-scaling the Lorenz curve in the following way:

$$LG_x(p) = \overline{x} \cdot L_x(p), \quad p \in [0,1], x \in D \tag{21}$$

where $L_x(p)$ stands for the Lorenz curve of the income vector x. Properties of these curves are easy to establish as direct consequences of Lorenz curves ones. Consequently, the dominance relationship can be established:

---

[q]It refers to the area located between the Lorenz curve and the diagonal of the unit square.

$$x \leq_{LG} y \Leftrightarrow LG_x(p) \geq LG_y(p), \ \forall p \in [0,1] \tag{22}$$

However, scale change induced through multiplication by its mean income implies that generalized curves don't measure inequality but they assume postulates related to social welfare valuation from a strictly monetary point of view[r]. Because of this reason, sometimes they are called *income-welfare curves* (Pena et al., 1996)[s].

Another interesting proposal has been the *ranks dominance criterion* (Nygard and Sandström, 1981), whose definition is presented next, assuming a pair of income vectors *x, y* ∈ D_N:

$$x \leq_R y \Leftrightarrow x_i \leq y_i, \ \forall i = 1,2,\ldots,N \tag{23}$$

closely related to majorization, as it can be observed. Again, this relation induces a partial order structure over D_N, as we may expect. Also, this relationship is related to the generalized Lorenz dominance, as it can be easily proved:

$$x \leq_R y \ \Rightarrow \ x \leq_{LG} y \tag{24}$$

Lately, a great deal of research effort has been devoted to the application of well-known *stochastic dominance criteria* to provide alternative tools in the study of economic inequality and other related concepts, such as poverty, welfare and so on[t]. Stochastic dominance consists of several relationships defined on pairs of random variables through their accumulative distribution functions. To define them, let X be a non-negative random variable, representing a society income and let F(.) be its accumulative distribution function, then the successive orders accumulative distribution functions can be defined through the following expressions:

$$\begin{aligned} F_1(z) &= F(z) = P(X \leq z), \ \forall z \geq 0 \\ F_j(z) &= \int_{-\infty}^{z} F_{j-1}(t) \cdot dt, \ \forall z \geq 0, \ \forall j = 2,3,\ldots. \end{aligned} \tag{25}$$

Furthermore, we can define the *j-order stochastic dominance criterion* as follows, where X, Y stands for two income non-negative random variables and F(.), G(.) are their respective cumulative distribution functions:

---

[r]Relations between Lorenz dominance and welfare have been studied in Bishop, Formby and Smith (1991) and subsequent papers.

[s]A sufficient condition for this dominance criterion is given in Ramos, Ollero and Sordo (2000).

[t]More details may be seen in Muliere and Scarsini (1989) or Bishop, Formby and Sakano (1995), among others.

$$x \leq_{D_j} y \iff F_j(z) \geq G_j(z), \quad \forall z \geq 0 \tag{26}$$

First and second orders of stochastic dominance criteria are strongly connected to rank and generalized Lorenz dominance relationships, respectively (Bishop, Formby and Smith, 1991). Again, all these criteria generate partial order structures, though there are progressively less non-comparable cases when the dominance order increases. At the same time, each order maintains the structure induced in inferior orders, so that if X dominates Y at the first order, then X will dominate Y at each order, for example. From a few years onward, research interest has placed on third-degree stochastic dominance to analyze its normative implications and what the decision would be if Lorenz curves crossed each other (Shorrocks and Foster, 1987; Davies and Hoy, 1994, 1995, among others).

It might be possible to define a total order structure by assuming the Rawls postulate (Rawls, 1972), and so focusing the comparison only on the poorer income. Thus, the *Rawls comparison criterion* would be defined as follows, where it is supposed to be $x, y \in$ D:

$$x \leq_{rw} y \iff \underset{i}{\mathrm{Min}}\{x_i\} \leq \underset{j}{\mathrm{Min}}\{y_j\} \tag{27}$$

But in this case, we are losing the sense of measuring inequality and what this criterion is comparing may be located nearer to poverty analysis.

In addition, there exist other more sophisticated proposals like successive orders Lorenz dominance criteria, but there is reasonable doubt about they might effectively measure inequality (Nygard and Sandström, 1981; Ramos and Sordo, 2001).

On the other hand, recently, absolute Lorenz curves have been proposed (Moyes, 1987), as an alternative. These curves are constructed using income differentials instead of classical relative ones, and so a new research field has emerged, where neither Pigou-Dalton Transfer Principle nor Scale Invariance Principle has to be included in their essential framework. Ramos and Sordo (2003) proved its relation to second-order absolute Lorenz ordering. Nevertheless, both this approach and generalized Lorenz curves are subject to the same conceptual controversy.

9. Inequality measures as an average of individual inequalities

In looking for Lorenz-compatible inequality measures, Theorem 7 and Corollary 1 allow us to consider economic inequality measures as averages of individual's income valuations. To understand this assertion, we would think of individual inequality as the amount each person contributes to global result with. In doing so, if the resources sharing was egalitarian (all individuals have perceived income mean), then each contribution to inequality would be null. But when some of them get more or less income than mean, they are contributing to raise inequality.

Therefore, the aim this interpretation is searching for consists in finding out the method we must use to measure such an individual contribution to inequality. It should be noted how this individual contribution must be coherent with inequality concepts, and so we might expect at least a reduction of optional indicators to choose among, as a result.

Next, along the first subsection, a precedent inequality indicators family addressed to this approach is exposed, whereas a new proposal about what an inequality indicator must fulfil will be presented at the second one.

9.1. *Generalized mean deviation family*

Castagnoli and Muliere (1991) consider inequality measures belonging to the following family:

$$C(x) = \sum_{i=1}^{N} \gamma_i \cdot |x_i - A| \ , \ x \in D_N ; \gamma_i \geq 0. \tag{28}$$

So, $\{\gamma_i, I = 1,2,\ldots,N\}$ stands for weights to averaging individual's inequality contributions valued by the function $|x_i - A|$, which expresses the differences between each income and A, as a reference point. Some particular cases are described below:

- Mean deviation related to income mean is obtained when $A = \mu$ and $\gamma_i = (1/N)$, $I = 1,2,\ldots,N$. With the same weightings, but $A = Me$, mean deviation with respect to median appears, where Me stands for median income.
- Pietra index is included too, using $A = \mu$ and $\gamma_i = 1/(2N\mu)$, $I = 1,2,\ldots,N$.
- Gini index, because it can be obtained using the following alternative expression (Berrebi and Silber, 1987):

$$I_G(x) = \frac{2}{N^2 \mu} \sum_{i=1}^{N} | N - 2i + 1 | \cdot | x_{N-i+1} - Me |, \quad x \in D_N \tag{29}$$

In addition, C(x) is an S-convex function if and only if the weights γ_i are non-increasing for $x_i < A$ and non-decreasing for $x_i > A$. In particular, this is true when all the weights are equal and positive.

A more general formulation consists of admitting the use of monotone non-decreasing functions g(.), so that:

$$C(x) = g^{-1}\left(\sum_{i=1}^{N} \gamma_i . g(|x_i - A|)\right), \ x \in D_N; \gamma_i \geq 0. \quad (30)$$

where $g^{-1}(t) = \inf\{x: g(x) \geq t\}$, including the first formulation when g(.) is the identity function.

This family values the individual inequality contribution through income differences respect to a reference point, usually mean or median income. Only the use of normalizing constants included in the weights specification, allows habitual relative indicators like Pietra and Gini ones can be obtained. Therefore, this family can be considered as generalized mean deviations.

9.2. *Individual inequality average indicators*

Corollary 1 shows a way of designing inequality measures through the characterization it contains. Therefore, separable convex functions defined on relative incomes are suitable tools to obtain "genuine" inequality indicators, by taking expectation over them. This is the justification of indicators as individual's inequality contribution averages, provided that they would be plenty of sense. Let us express these ideas through the definition below.

Definition 6: Let X be a non-negative random variable to model income and let $\mu = E(X)$ its expectation. Then an indicator I(.) is said to be an *individual inequality average* if it presents the next form:

$$I(X) = E\left[g\left(\frac{X}{\mu}\right)\right]$$

and the following conditions have to be fulfilled:

i) g(.) is a convex, continuous and real function.
ii) g(.) is non-negative.
iii) g(.) is non-increasing when $x < \mu$.
iv) g(.) is non-decreasing when $x > \mu$.

These conditions assure that I(X) will be a genuine inequality indicator, because they are imposing such a performance over the individual contribution valuation. As a matter of fact, the first one implies g(.) is a convex separable function, the second is necessary because individual contribution to inequality must not be negative, keeping in mind that incomes can not diminish inequality and only can accumulate it or not. Two last conditions allow us to impose the genuine perform of individual contributions, so as they must increase when income becomes more far away from mean, whatever the direction would be (egalitarian distribution represents absence of inequality, and then null individual contributions).

Gini index is one of the most renowned measures not included in this family, because it is a strictly S-convex function but not a convex separable one. Of course, this fact does not signify Gini index as a bad inequality index.

Nevertheless, generalized mean deviation family is included inside individual inequality average indices, when $A = \mu$ and $\{\gamma_i, I = 1,2,\ldots,N\}$ constitutes a probability distribution.

Next, some of the most usually proposed inequality measures are to be analyzed with relation to its belonging to this new family.

Proposition 2: Both Pietra index and squared coefficient of variation are individual inequality average indicators.

Proof

a) Pietra index can be expressed as:

$$I(X) = \frac{1}{2\mu}.E\big[|X-\mu|\big] = E\left[\frac{1}{2}\left|\frac{X}{\mu}-1\right|\right]$$

and so $g(x) = (\,|\,x-1\,|\,)/2$. Obviously, g(x) is a non-negative, continuous and real function. In addition, $g'(x) = (-1/2)$, if $x < 1$, and $g'(x) = 1/2$, when $x > 1$. Furthermore, $g''(x) = 0$ assures its convexity.

b) Squared coefficient of variation is:

$$CV^2(X) = \frac{1}{\mu^2}.E\left[(X-\mu)^2\right] = E\left[\left(\frac{X}{\mu}-1\right)^2\right]$$

then $g(x) = (x-1)^2$, and so it is a non-negative, continuous and real function.

Finally, $g'(x) = 2.(x-1)$ satisfies conditions iii) and iv), and $g''(x) = 2$ shows g(x) as a convex function.

Proposition 3: Both Theil order 1 and order 0 indicators are not individual inequality average indicators.

Proof

a) Theil order 1 index can be expressed as:

$$T_1(X) = E\left[\frac{X}{\mu}.\log\left(\frac{X}{\mu}\right)\right]$$

Hence, $g(x) = x.\log x$, and its first derivatives are $g'(x) = 1 + \log x$, $g''(x) = 1/x$.

So, this function is a convex, continuous and real one, but $g(x) \geq 0 \Leftrightarrow x \geq 1$, and it fails ii) because g(.) is negative when $x > 1$. This fact implies $T_1(X)$ admits negative contributions to inequality when incomes are lesser than mean. Also, condition iii) is not fulfilled.

b) Theil order 0 indicator is defined by:

$$T_0(X) = E\left[\log\left(\frac{\mu}{X}\right)\right] = E\left[\log\left(\frac{1}{X/\mu}\right)\right]$$

Hence, $g(x) = \log(1/x)$, $g'(x) = -(1/x)$, $g''(x) = 1/x^2$, and so it satisfies i).

But $g(x) \geq 0 \Leftrightarrow x < 1$, and it fails to satisfy ii). Then, $T_0(X)$ allows negative contributions in case incomes are greater than mean. In fact, condition iv) is not fulfilled.

Hence, Theil's indicators seem ill-conditioned to measure inequality, as far as it has proven in Proposition 3, taking into account implied reasons to fail[u]. So, the proposed family could be used to assure us about if convex separable inequality indicators are really measuring what they are supposed to do, despite its Lorenz-compatibility. Furthermore, this last result enlightens us about some well-known inequality measures, whose performance would not be adequate.

Perhaps, what this family shows is that Lorenz curves analyze inequality, of course, but it is possible other things may be included too. However, this is a task, which might require more investigation in the future.

10. Conclusions

In order to provide adequate comprehension of economic inequality measures, underlying statistical theory has been exposed along this paper. In doing so, we must conclude that economic inequality measures are firmly connected to majorization and Lorenz dominance relationships between pairs of income distributions. This conclusion has important consequences on the selection of inequality indicators to be compatible with those relationships.

To reach conclusions like the afore-mentioned, a historical revision has been developed to recover statistical terms related to economic inequality, seldom used nowadays, as well as its relations with currently trends and proposals. Even so, Lorenz curves have resisted against new theoretical approaches, since its appearance more than a century ago and they can be

[u] Dagum (1990, 2001) had warned about Theil,s indicators ill-conditioned performance, but he did it only in a social welfare framework.

considered as a cornerstone of inequality analysis, despite the great research efforts registered in this field.

Moreover, Pigou-Dalton Transfer Principle can be considered another cornerstone of inequality measures design, as it has been clearly shown in the paper. In fact, this property turns to be the essential underlying feature in majorization relations through the use of doubly stochastic matrices. This reason has conducted researchers toward the study of restrictions placed on this Principle, in order to investigate possible inequality indicator characterizations and to analyze concrete inequality measure performances, including their weighting schemes over income distributions.

Connection to Lorenz dominance criterion has been explained through Schur-convex, or S-convex, functions so that they are present in the construction of the most adequate kinds of indicators to measure inequality. As a matter of fact, convex separable functions and Karamata's Theorem are the key results in this line of work.

Basic inequality axioms are determined by relaxations carried out over the original majorization concept, which is defined only over pairs of equally-sized income vectors. So, Dalton Population Principle makes possible comparisons between different-sized income vectors, and Scale Invariance Axiom permits the same but using income vectors where the total amount of shared resources could be distinct. Foster's Theorem fixes what axioms are necessary to obtain Lorenz compatible inequality indicators. Additional restrictions like additive decomposition axioms are useful in characterizing inequality indicators families or in achieving interesting performances, but those properties rarely are linked to genuine inequality concepts.

Until now, other relations derived from global income vector comparisons have not allowed us to reach a total order structure over the income distributions space, and thus Lorenz dominance criterion continues as the most accepted background to measure economic inequality. This affirmation embodies Sen's statement about the partial order nature of economic inequality when income vector comparisons have to be considered and it suggests the use of batteries of economic indicators as a valid alternative.

Lorenz curves generating functions has revealed as an interesting approach to generate functional forms in the direct estimation of Lorenz curves task. Nevertheless, relations to density generating functions are hard to achieve, possibly because of the inherent difficult in establishing direct relations between Lorenz curves and cumulative distribution functions, except in the simplest cases. In addition, ordered families of Lorenz curves allow us to obtain a total order over income distributions, but it might be due to the assumption of scarcely realistic models of income.

Finally, obtaining a consensus inequality measure implies the necessity of more restrictions to be imposed together with the basic inequality axioms, but those new properties should not lose connection with the essential inequality concepts. Nevertheless, a set of properties has been stated in order to evaluate individual inequality contributions when convex separable functions are used as inequality indicators. Surprisingly, both Theil's order 0 and 1 fail to fulfill them, although they are Lorenz-compatible indicators. In the end, this fact might point out towards Lorenz curves measure inequality, of course, but other things may be involved too.

Acknowledgments

The author gratefully acknowledges partial financial support from University of Alcalá (grant UAH-PI2004/034) and of Junta de Comunidades de Castilla-La Mancha together with Fondo Social Europeo. (Project PBI-05-004).

References

1. B.C. Arnold. (1987). Majorization and the Lorenz Order: A Brief Introduction. Lecture Notes in Statistics. New York: Springer Verlag.
2. B.C. Arnold. (2005). The Lorenz curve: Evergreen after 100 years. Int. Conference in Memory of C. Gini and M.O. Lorenz. Siena. [http://www.unisi.it/eventi/GiniLorenz05].
3. B.C. Arnold, C.A. Robertson, P.L. Brockett and B.Y. Shu. (1987). Generating ordered families of Lorenz curves by strongly unimodal distributions. Journal of Business and Economic Statistics, 5(2), 305–308.
4. A.B. Atkinson. (1970). On the measurement of inequality. Journal of Economic Theory, 2, 244–263.
5. C.P.A. Bartels. (1977). Economics Aspects of Regional Welfare. Martinus Nijhoff Sciences Division.
6. C.M. Beach and R. Davidson. (1983). Distribution-free statistical inference with Lorenz curves and income shares. Review of Economic Studies, L, 723–735.
7. C.M. Beach and J. Richmond. (1985). Joint confidence intervals for income shares and Lorenz curves. International Economic Review, 26(2), 439–450.
8. Z.M. Berrebi and J. Silber. (1987). Dispersion, asymmetry and the Gini index of inequality. International Economic Review, 28(2), 331–338.
9. G. Birkhoff. (1976). Tres observaciones sobre el Álgebra Lineal. Univ. Nacional de Tucuman Rev., Serie A, 5, 147–151.

10. J.A. Bishop, J.P. Formby and R. Sakano. (1995). Lorenz and stochastic-dominance comparisons of European income distributions. Research on Economic Inequality, 6, 77–92.
11. J.A. Bishop, J.P. Formby and W.J. Smith. (1991). Lorenz dominance and welfare: Changes in the U.S. distribution of income, 1967–1986. Review of Economics and Statistics, 73, 134–139.
12. F. Bourguignon. (1979). Decomposable income inequality measures. Econometrica, 47, 901–920.
13. J. Callejón. Un nuevo método para generar distribuciones de probabilidad. Problemas asociados y aplicaciones. Ph. D. dissertation. University of Granada.
14. J.M. Casas, R. Herrerías and J.J. Núñez. (1997). Familias de Formas Funcionales para estimar la Curva de Lorenz. Actas de la IV Reunión Anual de ASEPELT-España. Servicio de Estudios de Cajamurcia, 171–176. Reprinted in Aplicaciones estadísticas y económicas de los sistemas de funciones indicadoras (R. Herrerías, F. Palacios and J. Callejón, eds.). Univ. Granada, 119–125 (2001).
15. J.M. Casas and J.J. Núñez. (1987). Algunas Consideraciones sobre las Medidas de Concentración. Aplicaciones. Actas de las II Jornadas sobre Modelización Económica, 49–62. Barcelona. Reprinted in Aplicaciones estadísticas y económicas de los sistemas de funciones indicadoras (R. Herrerías, F. Palacios and J. Callejón, eds.). Univ. Granada, 111–118 (2001).
16. J.M. Casas and J.J. Núñez. (1991). Sobre la Medición de la Desigualdad y Conceptos Afines. Actas de la V Reunión Anual de ASEPELT-España, Caja de Canarias, 2, 77–84. Reprinted in Aplicaciones estadísticas y económicas de los sistemas de funciones indicadoras (R. Herrerías, F. Palacios and J. Callejón, eds.). Univ. Granada, 127–133 (2001).
17. E. Castagnoli and P. Muliere. (1990). A note on inequality measures and the Pigou-Dalton Principle of Transfers. Income and Wealth Distribution, Inequality and Poverty. (C. Dagum and M. Zenga, eds.) Springer Verlag, 171–127.
18. E. Castillo, A.S. Hadi and J.M. Sarabia. (1998). A method for estimating Lorenz curves. Communications in Statistics, Theory and Methods, 27, 2037–2063.
19. F.A. Cowell, Measuring inequality. 2ª ed. LSE Handbooks in Economics. Prentice Hall/Harvester Wheatsheaf (1995).
20. C. Dagum. (1990). Relationship between income inequality measures and social welfare functions. Journal of Econometrics, 43(1–2), 91–102.
21. C. Dagum. (2001). Desigualdad del rédito y bienestar social, descomposición, distancia direccional y distancia métrica entre distribuciones. Estudios de Economía Aplicada, 17, 5–52.
22. H. Dalton. (1920). The measurement of the inequality of incomes. Economic Journal, 30, 348–361.

23. J. Davies and M. Hoy. (1994). The normative significance of using third-degree stochastic dominance in comparing income distributions. Journal of Economic Theory, 64, 520–530.
24. J. Davies and M. Hoy. (1995). Making inequality comparisons when Lorenz curves intersect. American Economic Review, 85(4), 980–986.
25. J. Domínguez and J.J. Núñez. (2005). The evolution of economic inequality in the EU countries during the nineties. First Meeting of the Society for the Study of Economic Inequality (ECINEQ). Palma de Mallorca. Available at [http://www.ecineq.org]
26. M. Fleurbaey and P. Michel. (2001). Transfer Principles and inequality aversion, with an application to optimal growth. Mathematical Social Sciences, 42, 1–11.
27. J.E. Foster. (1985). Inequality measurement. Published in Fair Allocation (H.P. Young, ed.), Proceedings of Symposia in Applied Mathematics, 33, Providence, American Mathematical Society, 31–68.
28. C. García, J.J. Núñez, L.F. Rivera and A.I. Zamora. (2002). Análisis comparativo de la desigualdad a partir de una batería de indicadores. El caso de las Comunidades Autónomas españolas en el periodo 1973–1991. Estudios de Economía Aplicada, 20(1), 137–154.
29. R.M. García and J.M. Herrerías. (2001). Inclusión de curvas de Lorenz en las funciones generadoras. Aplicaciones estadísticas y económicas de los sistemas de funciones indicadoras (R. Herrerías, F. Palacios and J. Callejón, eds.). Univ. Granada, 185–191.
30. J.L. Gastwirth. (1971). A general definition of the Lorenz curve. Econometrica, 39, 1037–1039.
31. C. Gini. (1912). Variabilitá e Mutabilitá: Contributo allo studio delle distribuzioni e relazioni statistiche. Studi Economico-Giuridici dell'Universitá di Cagliari, 3, 1–158.
32. C. Gini. (1921). Measurement of inequality of incomes. The Economic Journal, 31, 124–126.
33. C.M. Goldie. (1977). Convergence Theorems for empirical Lorenz curves and their inverses. Advances in Applied Probability, 9, 765–791.
34. M.R. Gupta. (1984). Functional form for estimating the Lorenz curve. Econometrica, 52(5), 1313–1314.
35. G.H. Hardy, J.E. Littlewood and G. Polya. (1929). Some simple inequalities satisfied by convex functions. The Messenger of Mathematics, 26, 145–153.
36. G.H. Hardy, J.E. Littlewood and G. Polya. (1952). Inequalities. 2ª ed. Cambridge University Press.
37. R. Herrerías, F. Palacios and J. Callejón. (2001). Las curvas de Lorenz y el sistema de Pearson. Published in Aplicaciones estadísticas y económicas de los sistemas de funciones indicadoras (R. Herrerías, F. Palacios and J. Callejón, eds.). Univ. Granada, 135–151.
38. J.C. Houghton. (1978). Birth of a parent: The Wakeby distribution for modelling flood flows. Water Resources Research, 14, 1105–1109.

39. J. Iritani and K. Kuga. (1983). Duality between the Lorenz curves and the income distribution functions. Economic Studies Quarterly, 23, 9–21.
40. N.C. Kakwani. (1980). Income Inequality and Poverty. Methods of Estimation and Policy Applications. Oxford University Press.
41. N.C. Kakwani and N. Podder. (1973). On the estimation of Lorenz curves from grouped observations. International Economic Review, 14(2), 278–291.
42. J. Karamata. (1932). Sur une inégalité rélative aux fonctions convexes. Publ. Math. Univ. Belgrade, 1, 145–148.
43. M. Kendall and A. Stuart. (1977). The Advanced Theory of Statistics, 1, 4ª ed. C. Griffin. London.
44. S. Kuznets. (1953). Share of upper income groups in income and savings. National Bureau of Economic Research. New York.
45. M.O. Lorenz. (1905). Methods of measuring the concentration of wealth. Journal of the American Statistical Association, 9, 209–219.
46. A.W. Marshall and I. Olkin, Inequalities: Theory of Majorization and its Applications. New York: Academic Press.
47. F. Mehran. (1976). Linear measures of income inequality. Econometrica, 44, 805–809.
48. P. Moyes. (1987). A new concept of Lorenz domination. Economics Letters, 23, 203–207.
49. R.F. Muirhead. (1903). Some methods applicable to identities and inequalities of symmetric algebraic functions of n letters. Proceedings of Edinburgh Mathematical Society, 21, 144–157.
50. P. Muliere and M. Scarsini. (1989). A note on stochastic dominance and inequality measures. Journal of Economic Theory, 49, 314–323.
51. F. Nygard and A. Sandström. (1981). Measuring Income Inequality. Stockholm: Amqvist and Wiksell International.
52. A.M. Ostrowski. (1952). Sur quelques applications des fonctions convexes et concaves au sens de I. Schur. Journal of Math. Pures Appl., 9, 253–292.
53. V. Pareto. (1897). Cours d'Economie Politique. Rouge. Lausanne.
54. J.B. Pena (Dir.), F.J. Callealta, J.M. Casas, A. Merediz and J.J. Núñez. (1996). Distribución Personal de la Renta en España. Pirámide. Madrid.
55. G. Pietra. (1914–15). Delle relazioni tra gli indici di variabilitá. Note I in Atti del R. Istituto Veneto di Scienze, Lettere ed Arti, LXXIV (II), 775–804.
56. G. Pietra. (1948). Studi di statistica metodologica. Giuffre. Milan.
57. A.C. Pigou. (1912). Wealth and welfare. McMillan. New York.
58. J.S. Ramberg, E.J. Dudewicz, P.R. Tadikamalla and E.F. Mykytra. (1979). A probability distribution and its uses in fitting data. Technometrics, 21, 201–214.
59. H.M. Ramos, J. Ollero and MA. Sordo. (2000). A sufficient condition for generalizad Lorenz order. Journal of Economic Theory, 90, 286–292.

60. H.M. Ramos and M.A. Sordo. (2001). El orden de Lorenz generalizado de orden j, ¿un orden en desigualdad?. Estudios de Economía Aplicada, 19, 139–149.
61. H.M. Ramos and M.A. Sordo. (2003). Dispersion measures and dispersive orderings. Statistics and Probability Letters, 61, 123–131.
62. J. Rawls. (1972). A Theory of Justice. London: Oxford University Press.
63. J. Ruiz-Castillo. (1986). Problemas conceptuales en la medición de la desigualdad. Hacienda Pública Española, 101, 17–31.
64. J. Ruiz-Castillo. (1987). La medición de la pobreza y de la desigualdad en España, 1980-81. Estudios Económicos, 42. Servicio de Estudios del Banco de España. Madrid.
65. J.M. Sarabia, E. Castillo and D. Slottje. (1999). An ordered family of Lorenz curves. Journal of Econometrics, 91, 43–60.
66. J.M. Sarabia, E. Castillo and D. Slottje. (2002). Lorenz ordering between McDonald's generalized functions of the income size distribution. Economic Letters. 75, 265–270.
67. I. Schur. (1923). Uber eine klasse von mittelbildungen mit anwendungen die determinaten. Theorie Sitzungsber Berlin Math. Gesellschaft, 22, 9–20.
68. R.R. Schutz. (1951). On the measurement of income inequality. American Economic Review, 41, 107–122.
69. A.K. Sen. (1973). On Economic Inequality. Oxford: Clarendon Press.
70. A.K. Sen and J.E. Foster. (1997). On Economic Inequality. Expanded edition. Clarendon Press Paperbacks and Oxford University Press.
71. A. Shorrocks. (1983). Ranking income distributions. Economica, 50, 3–18.
72. A. Shorrocks and J.E. Foster. (1987). Transfer sensitive inequality measures. Review of Economic Studies, 54, 485–497.
73. H. Wold. (1935). A study of the mean difference, concentration curves and concentration ratio. Metron, 12, 39–58.
74. I. Zubiri. (1985). Una introducción al problema de la medición de la desigualdad. Hacienda Pública Española, 95, 291–317.

Chapter 12

EXTENDED WARING BIVARIATE DISTRIBUTION

J. RODRÍGUEZ-AVI

Department of Statistics and Operations Research, University of Jaén
Campus Las Lagunillas, B3, Jaén, 23071, Spain

A. CONDE-SÁNCHEZ

Department of Statistics and Operations Research, University of Jaén
Campus Las Lagunillas, B3, Jaén, 23071, Spain

A.J. SÁEZ-CASTILLO

Department of Statistics and Operations Research, University of Jaén
Campus Las Lagunillas, B3, Jaén, 23071, Spain

M.J. OLMO-JIMÉNEZ

Department of Statistics and Operations Research, University of Jaén
Campus Las Lagunillas, B3, Jaén, 23071, Spain

The aim of this paper is to obtain a bivariate distribution that extends the Bivariate generalized Waring distribution (*BGWD*) and that preserves some of its properties, such as the partition of the variance into three distinguishable components due to randomness, proneness and liability. Finally, an example in the context of accident theory is included in order to illustrate the versatility of this new distribution.

1. Introduction

Accident theory has become the object of numerous studies that tried to develop several hypotheses in order to interpret the causes of an accident. Among them, the idea of accident proneness has stimulated much interesting statistical theories. One important contribution in this direction is the "proneness-liability" model proposed by Irwing [1] and Xekalaki [5] giving rise to a three parameter discrete distribution, the univariate generalized Waring distribution (*UGWD*) with probability generating function (p.g.f.) given by the Gauss hipergeometric function:

$$G(t) = \frac{(\rho)_k}{(a+\rho)_k} \, {}_2F_1(a,k;a+k+\rho;t), \tag{1}$$

where a, k, $\rho>0$.

This model assumes that all non-random factors may be split into internal and external factors. So, the term "accident proneness" refers to a person's predisposition to accidents and the term "accident liability" refers to a person's exposure to external risk of accident. Then, the *UGWD* arises from a Poisson distribution where the parameter λ is the "liability" that follows a Gamma distribution and the parameter p is the "proneness" that follows a Beta distribution, that is:

$$Poisson(\Lambda) \underset{\Lambda}{\wedge} Gamma\left(a, \frac{1-P}{P}\right) \underset{P}{\wedge} Beta\ I(\rho,k). \tag{2}$$

This way of obtaining the distribution as a mixture allows the variability to be split into three additive components due to proneness, liability and randomness:

$$Var(X) = \underbrace{\frac{ak}{\rho-1}}_{\text{randomness}} + \underbrace{\frac{ak(k+1)}{(\rho-1)(\rho-2)}}_{\text{liability}} + \underbrace{\frac{a^2k(\rho+k-1)}{(\rho-1)^2(\rho-2)}}_{\text{proneness}}. \tag{3}$$

However, there is a problem arising from the fact that the *UGWD* is symmetrical in the parameters a and k and, hence, distinguishable estimates for non-random components cannot be obtained.

Moreover, it is observed that the *UGWD* belongs to the family of Gaussian hypergeometric distributions, *GHD* (Kemp and Kemp [2]). Thus, Rodríguez *et al.* [4] have considered an extension of this distribution, introducing a parameter λ, $0<\lambda\leq 1$, in such a way that the p.g.f. is given by:

$$G(t) = \frac{{}_2F_1(\alpha,\beta;\gamma;\lambda t)}{{}_2F_1(\alpha,\beta;\gamma;\lambda)} \qquad \alpha,\beta,\gamma>0, \quad 0<\lambda\leq 1. \tag{4}$$

This distribution, denoted by $GHDI(\alpha,\beta,\gamma,\lambda)$, may also be obtained as a mixture of a Poisson distribution with a Gamma and a generalized Beta distributions, so that the property of partition of the variance is verified and data that can not be adequately fitted by the *UGWD*, are successfully modeled by the proposed distribution. However, the two non-random variance components cannot be separately estimated either.

Xekalaki [6] proposed a solution of this problem dividing the whole period of observation into two non-overlapping sub-periods and then studying the resulting bivariate accident distribution. Following a similar process to the univariate case, this distribution, that she called bivariate generalized Waring distribution (*BGWD*), has p.g.f. generated by the F_1 Appell's hypergeometric function:

$$G(t_1,t_2)=\frac{(\rho)_{k+m}}{(a+\rho)_{k+m}}F_1(a;k,m;a+k+m+\rho;t_1,t_2), \tag{5}$$

where

$$F_1(\alpha;\beta_1,\beta_2;\gamma;\lambda_1,\lambda_2)=\sum_{x=0}^{\infty}\sum_{y=0}^{\infty}\frac{(\alpha)_{x+y}(\beta_1)_x(\beta_2)_y\lambda_1^x\lambda_2^y}{(\gamma)_{x+y}x!y!}, \tag{6}$$

with a, k, m, $\rho>0$.

Then, the accident distribution in the whole period is also a *UGWD*, like in each one of the sub-periods considered. Moreover, in this situation it is possible to distinguish the non-random components in the partition of the variance. In Kocherlakota and Kocherlakota [3] some of the most interesting properties of the *UGWD* are listed.

Our aim is to obtain a bivariate distribution that extends the *BGWD* introducing a parameter λ, but without loosing its excellent properties in order to be used in fields such as accident theory. Thus, distinguishable estimates for the two non-random variance components are obtained and, moreover, fits achieved by the *BGWD* are improved.

2. Extension of the *BGWD* in accident theory

We will generalize the result obtained by Xekalaki [6] that presents the bivariate Waring distribution as a mixture of a double Poisson distribution with two independent Gamma distributions and a Beta distribution.

We consider that the number of accidents that a person incurs in two consecutive sub-periods is determined by a proneness (internal risk), constant throughout the entire period of observation, and by a liability (external risk) that varies from one period to the other. This hypothesis seems to be reasonable, at least for a limited period of time, as Xekalaki points out. In this situation, let $(X,Y,\Lambda_1,\Lambda_2,P)$ be a random vector where $\Lambda_1|P$ and $\Lambda_2|P$ represent liability in each period and P proneness, so that:

- $(X,Y)|\Lambda_1=l_1,\Lambda_2=l_2,P=p$ has a double Poisson distribution with probability mass function (p.m.f.)

$$f_{(x,y)|\Lambda_1=l_1,\Lambda_2=l_2,P=p}(x,y)=e^{-l_1}\frac{l_1^x}{x!}e^{-l_2}\frac{l_2^y}{y!}. \tag{7}$$

This means that the number of accidents in each period has a Poisson distribution, both independent.

- Liability parameters have two independent Gamma distributions:

$$\begin{aligned} \Lambda_1|_{P=p} &\to Gamma(\beta_1, \upsilon) \\ \Lambda_2|_{P=p} &\to Gamma(\beta_2, \upsilon), \end{aligned} \tag{8}$$

with $\nu=\lambda(1-p)/[1-\lambda(1-p)]$, β_1, $\beta_2>0$ and density function

$$\frac{l_i^{\beta_i-1} e^{-l_i/\upsilon}}{\Gamma(\beta_i)\upsilon^{\beta_i}}. \tag{9}$$

- P has a generalized Beta distribution with density

$$f_P(p) = \frac{1}{F_1(\alpha;\beta_1,\beta_2;\gamma;\lambda,\lambda)} \frac{\Gamma(\gamma)}{\Gamma(\alpha)\Gamma(\gamma-\alpha)} \frac{p^{\gamma-\alpha-1}(1-p)^{\alpha-1}}{(1-\lambda(1-p))^{\beta_1+\beta_2}}, \tag{10}$$

where $\gamma>\alpha$, $0<\lambda\leq 1$ and $0<p<1$.

Therefore:

1. $(X,Y)|P=p$ has a double negative Binomial distribution with m.p.f.:

$$f_{(X,Y)|_{P=p}}(x,y) = \frac{(\beta_1)_x}{x!}\frac{(\beta_2)_y}{y!}(1-\lambda(1-p))^{\beta_1+\beta_2}(\lambda(1-p))^{x+y}. \tag{11}$$

2. (X,Y) is an extended bivariate Waring distribution (from now on *EBWD*) with p.m.f.:

$$f_{(X,Y)}(x,y) = f_0 \frac{(\alpha)_{x+y}(\beta_1)_x(\beta_2)_y \lambda^{x+y}}{(\gamma)_{x+y} x! y!}, \tag{12}$$

where the constant of normalization, f_0, is

$$f_0 = F_1(\alpha;\beta_1;\beta_2;\gamma;\lambda,\lambda)^{-1} = {}_2F_1(\alpha;\beta_1+\beta_2;\gamma;\lambda)^{-1}. \tag{13}$$

Below, we are going to prove these statements:

1. Integrating in l_1 and l_2:

$$
\begin{aligned}
&f_{(X,Y)|P=p}(x,y) \\
&= \int_0^\infty \int_0^\infty e^{-l_1} \frac{l_1^x}{x!} e^{-l_2} \frac{l_2^y}{y!} \frac{l_1^{\beta_1 - 1} e^{-l_1/\upsilon}}{\Gamma(\beta_1)\upsilon^{\beta_1}} \frac{l_2^{\beta_2 - 1} e^{-l_2/\upsilon}}{\Gamma(\beta_2)\upsilon^{\beta_2}} dl_1 dl_2 \\
&= \frac{1}{x!\,y!} \frac{1}{\Gamma(\beta_1)\Gamma(\beta_2)\upsilon^{\beta_1+\beta_2}} \int_0^\infty e^{-l_1(1+\upsilon^{-1})} l_1^{x+\beta_1-1} dl_1 \\
&\quad \cdot \int_0^\infty e^{-l_2(1+\upsilon^{-1})} l_2^{y+\beta_2-1} dl_2 \\
&= \frac{1}{x!\,y!} \frac{1}{\Gamma(\beta_1)\Gamma(\beta_2)\upsilon^{\beta_1+\beta_2}} \Gamma(x+\beta_1)\Gamma(y+\beta_2) \left(1+\frac{1}{\upsilon}\right)^{-(x+\beta_1+y+\beta_2)} \qquad (14) \\
&= \frac{(\beta_1)_x(\beta_2)_y}{x!\,y!} \left(\frac{1}{\upsilon}\right)^{\beta_1+\beta_2} \left(\frac{\upsilon}{\upsilon+1}\right)^{x+y+\beta_1+\beta_2} \\
&= \frac{(\beta_1)_x(\beta_2)_y}{x!\,y!} \left(\frac{1}{\upsilon+1}\right)^{\beta_1+\beta_2} \left(\frac{\upsilon}{\upsilon+1}\right)^{x+y} \\
&= \frac{(\beta_1)_x(\beta_2)_y}{x!\,y!} (1-\lambda(1-p))^{\beta_1+\beta_2} (\lambda(1-p))^{x+y}.
\end{aligned}
$$

2. Firstly, we note that since

$$
F_1(\alpha;\beta_1,\beta_2;\gamma;\lambda,\lambda) = \frac{\Gamma(\gamma)}{\Gamma(\alpha)\Gamma(\gamma-\alpha)} \int_0^1 \frac{p^{\gamma-\alpha-1}(1-p)^{\alpha-1}}{(1-\lambda(1-p))^{\beta_1+\beta_2}} dp \qquad (15)
$$

the function in Eq. (10) is a density one. Then,

$$
\begin{aligned}
f_{(X,Y)}(x,y) &= \int_0^1 \frac{(\beta_1)_x(\beta_2)_y}{x!\,y!} \lambda^{x+y}(1-p)^{x+y} \frac{1}{F_1(\alpha;\beta_1,\beta_2;\gamma;\lambda,\lambda)} \\
&\quad \times \frac{\Gamma(\gamma)}{\Gamma(\alpha)\Gamma(\gamma-\alpha)} p^{\gamma-\alpha-1}(1-p)^{\alpha-1} dp \\
&= \frac{(\beta_1)_x(\beta_2)_y}{x!\,y!} \frac{\lambda^{x+y}}{F_1(\alpha;\beta_1,\beta_2;\gamma;\lambda,\lambda)} \frac{\Gamma(\gamma)}{\Gamma(\alpha)\Gamma(\gamma-\alpha)} \\
&\quad \times \int_0^1 p^{\gamma-\alpha-1}(1-p)^{x+y+\alpha-1} dp \qquad (16) \\
&= \frac{(\beta_1)_x(\beta_2)_y}{x!\,y!} \frac{\lambda^{x+y}}{F_1(\alpha;\beta_1,\beta_2;\gamma;\lambda,\lambda)} \frac{\Gamma(\gamma)}{\Gamma(\alpha)\Gamma(\gamma-\alpha)} \\
&\quad \times \frac{\Gamma(\gamma\quad\alpha)\Gamma(x+y+\alpha)}{\Gamma(x+y+\gamma)} \\
&= \frac{1}{F_1(\alpha;\beta_1,\beta_2;\gamma;\lambda,\lambda)} \frac{(\alpha)_{x+y}(\beta_1)_x(\beta_2)_y\lambda^{x+y}}{(\gamma)_{x+y}\,x!\,y!}.
\end{aligned}
$$

It can be observed that if $\lambda=1$ the expressions in Eqs. (10), (11), (12) and (13) reduce to those deduced by Xekalaki [4].

3. Properties of the *EBWD*

In this section we show some of the properties of the *EBWD*. Firstly, the p.g.f. is given by:

$$g(t_1,t_2) = f_0 F_1(\alpha,\beta_1,\beta_2;\gamma;\lambda t_1,\lambda t_2), \tag{17}$$

which is convergent for $|t_1|\leq 1$, $|t_2|\leq 1$ if $|\lambda|\leq 1$ (with $\gamma>\alpha+\beta_1+\beta_2$ for the case in that $\lambda=1$).

The probabilities may be obtained in a recursive way, since this distribution, like the *BGWD*, belongs to the Pearson's system. Then, the p.m.f., $f_{r,s}$, satisfies the following system of difference equations:

$$\begin{aligned}(\gamma+r+s)(r+1)f_{r+1,s}-\lambda(\alpha+r+s)(\beta_1+r)f_{r,s}&=0\\ (\gamma+r+s)(s+1)f_{r,s+1}-\lambda(\alpha+r+s)(\beta_2+s)f_{r,s}&=0.\end{aligned} \tag{18}$$

So, if the constant of normalization, $f_{0,0}=f_0$ given in Eq. (13), is known, the remainder probabilities are obtained. When $\lambda=1$ this constant may be computed exactly from the Gauss summation theorem:

$$\begin{aligned}F_1(\alpha,\beta_1,\beta_2;\gamma;1,1)&=\frac{(\gamma-\beta_1-\beta_2)_{\beta_1+\beta_2}}{(\gamma-\alpha-\beta_1-\beta_2)_{\beta_1+\beta_2}}\\ &=\frac{\Gamma(\gamma)\Gamma(\gamma-\alpha-\beta_1-\beta_2)}{\Gamma(\gamma-\beta_1-\beta_2)\Gamma(\gamma-\alpha)}.\end{aligned} \tag{19}$$

In the general case, the value of this constant is computed by approximation.

3.1. *Mixture of bivariate confluent hypergeometric distributions*

Xekalaki [7] proves that the *EBWD* may be obtained as a mixture of a generalized Gamma distribution and a bivariate confluent hypergeometric distribution. Specifically, suppose that:

- $(X,Y)|\Lambda=l$ has a joint distribution with p.g.f. given by:

$$g(t_1,t_2)=\frac{\phi_2(\beta_1,\beta_2;\gamma;lt_1,lt_2)}{{}_1F_1(\beta_1+\beta_2;\gamma;l)}, \tag{20}$$

where

$$\phi_2(a,b;c;x,y)+\sum_{i=0}^{\infty}\sum_{j=0}^{\infty}\frac{(a)_i(b)_j}{(c)_{i+j}}\frac{x^i y^j}{i!j!}. \tag{21}$$

- Λ has a generalized Gamma distribution with density given by:

$$f(l)=\frac{{}_1F_1(\beta_1+\beta_2;\gamma;l)}{\lambda^{\alpha}\Gamma(\alpha)_2F_1(\alpha,\beta_1+\beta_2;\gamma;\lambda)}l^{\alpha-1}e^{-l/\lambda}. \tag{22}$$

Then, the p.g.f. of (X,Y) is:

$$G(t_1,t_2)=\frac{F_1(\alpha,\beta_1,\beta_2;\gamma;\lambda t_1,\lambda t_2)}{{}_2F_1(\alpha,\beta_1+\beta_2;\gamma;\lambda)}. \tag{23}$$

However, Xekalaki does not study this distribution in depth.

3.2. *Marginal and conditional distributions*

The marginal distributions for $\lambda=1$ are generated by a ${}_2F_1(\alpha,\beta_1;\gamma-\beta_2;1)$ and a ${}_2F_1(\alpha,\beta_2;\gamma-\beta_1;1)$, respectively. Therefore, they are *UGWD*.

The following result is verified for any λ:

$$\begin{aligned} f_r &= f_0\sum_{s=0}^{\infty}\frac{(\alpha)_{r+s}(\beta_1)_r(\beta_2)_s}{(\gamma)_{r+s}}\frac{\lambda^r\lambda^s}{r!s!} \\ &= f_0\frac{(\alpha)_r(\beta_1)_r}{(\gamma)_r}\frac{\lambda^r}{r!}\sum_{s=0}^{\infty}\frac{(\alpha+r)_s(\beta_2)_s}{(\gamma+r)_s}\frac{\lambda^s}{s!} \\ &= f_0\frac{(\alpha)_r(\beta_1)_r}{(\gamma)_r}\frac{\lambda^r}{r!}\,{}_2F_1(\alpha+r,\beta_2;\gamma+r;\lambda). \end{aligned} \tag{24}$$

Thus, the marginal distributions have the p.m.f.:

$$\begin{aligned} f_r &= f_0\frac{(\alpha)_r(\beta_1)_r\lambda^r}{(\gamma)_r r!}\,{}_2F_1(\alpha+r,\beta_2;\gamma+r;\lambda) \\ f_s &= f_0\frac{(\alpha)_s(\beta_2)_s\lambda^s}{(\gamma)_s s!}\,{}_2F_1(\alpha+s,\beta_1;\gamma+s;\lambda), \end{aligned} \tag{25}$$

where f_0 is the constant of normalization given in Eq. (13). Then, it should be emphasized that:

- The marginal distributions are not *GHD*, but they are *UGWD* when $\lambda=1$, so they are more general distributions that the Waring distribution.
- We may obtain the p.g.f. of the marginal distributions since:

$$\begin{aligned} g_X(t) = g(t,1) &= \frac{F_1(\alpha,\beta_1,\beta_2;\gamma;\lambda t,\lambda)}{{}_2F_1(\alpha,\beta_1+\beta_2;\gamma;\lambda)} \\ g_Y(t) = g(1,t) &= \frac{F_1(\alpha,\beta_1,\beta_2;\gamma;\lambda,\lambda t)}{{}_2F_1(\alpha,\beta_1+\beta_2;\gamma;\lambda,\lambda)} \end{aligned} \quad (26)$$

Another important question that will be finalized later, is the distribution of $X+Y$, that is, the distribution of the number of accident in the whole period:

$$g_{X+Y}(t) = g(t,t) = \frac{F_1(\alpha,\beta_1,\beta_2;\gamma;\lambda t,\lambda t)}{F_1(\alpha,\beta_1,\beta_2;\gamma;\lambda,\lambda)} = \frac{{}_2F_1(\alpha,\beta_1+\beta_2;\gamma;\lambda t)}{{}_2F_1(\alpha,\beta_1+\beta_2;\gamma;\lambda)}, \quad (27)$$

It is a GHDI$(\alpha,\beta_1+\beta_2,\gamma,\lambda)$, as it was desirable. Hence, the total number of accidents has a *GHD*I, independently of the division in two sub-periods, while the number of accidents has a distribution with p.m.f given by Eq. (25) in each sub-period.

In order to obtain the conditional distributions, we can operate in the following way:

$$f_{r/s} = \frac{f_{r,s}}{f_s} = \frac{f_0 \dfrac{(\alpha)_{r+s}(\beta_1)_r(\beta_2)_s}{(\gamma)_{r+s}} \dfrac{\lambda^r\lambda^s}{r!s!}}{f_0 \dfrac{(\alpha)_s(\beta_2)_s\lambda^s}{(\gamma)_s s!}\, {}_2F_1(\alpha+s,\beta_1;\gamma+s;\lambda)}, \quad (28)$$

having the expressions

$$\begin{aligned} f_{r/s} &= f_{0/s}\frac{(\alpha+s)_r(\beta_1)_r\lambda^r}{(\gamma+s)_r r!} \\ f_{s/r} &= f_{0/r}\frac{(\alpha+r)_s(\beta_2)_s\lambda^s}{(\gamma+r)_s s!}, \end{aligned} \quad (29)$$

where $f_{0/s}={}_2F_1(\alpha+s,\beta_1;\gamma+s;\lambda)^{-1}$ and $f_{0/r}={}_2F_1(\alpha+r,\beta_2;\gamma+r;\lambda)^{-1}$.

Their p.g.f., therefore, are:

$$\begin{aligned} g(t) &= f_{0/s}\,{}_2F_1(\alpha+s,\beta_1;\gamma+s;\lambda t) \\ g(t) &= f_{0/r}\,{}_2F_1(\alpha+r,\beta_2;\gamma+r;\lambda t). \end{aligned} \quad (30)$$

So, these distributions belong to the *GHD*I family.

3.3. *Components of the variance*

Xekalaki [6] obtained the components of the variance for $X+Y$ that, in our case, has a *GHD*I. So, we have the following variance components (Rodríguez *et al.* [4]):

$$\sigma^2 = Var(X+Y) = \underbrace{(\beta_1+\beta_2)E_P(V)}_{\text{randomness}} + \underbrace{(\beta_1+\beta_2)E_P(V^2)}_{\text{liability}} + \underbrace{(\beta_1+\beta_2)^2 Var_P(V)}_{\text{proneness}}, \tag{31}$$

where $V=\lambda(1-P)/[1-\lambda(1-P)]$ and P has a distribution with the density function given in Eq. (10).

Concerning X and Y, since both variables are obtained as mixtures, their variances may be split into three components

$$\begin{aligned}\sigma_X^2 &= Var(X) = \beta_1 E_P(V) + \beta_1 E_P(V^2) + \beta_1^2 Var_P(V)\\ \sigma_Y^2 &= Var(Y) = \beta_2 E_P(V) + \beta_2 E_P(V^2) + \beta_2^2 Var_P(V),\end{aligned} \tag{32}$$

in the same way as the *BGWD*.

4. Applications

To conclude, we consider data about the number of driver accidents in Connecticut (Xekalaki [6]).

The parameters are estimated by the maximum likelihood method because the method of moments does not provide good estimates. Then, the log-likelihood function, whose expression is

$$\begin{aligned}\ln L(\alpha,\beta_1,\beta_2,\gamma,\lambda) &= n\ln f_0 + \sum_{i=1}^{n}\ln(\alpha)_{x_i+y_i} + \sum_{i=1}^{n}\ln(\beta_1)_{x_i}\\ &+\sum_{i=1}^{n}\ln(\beta_2)_{y_i} - \sum_{i=1}^{n}\ln(\gamma)_{x_i+y_i}\\ &+\ln\lambda\sum_{i=1}^{n}(x_i+y_i) - \sum_{i=1}^{n}\ln x_i! - \sum_{i=1}^{n}\ln y_i!,\end{aligned} \tag{33}$$

is maximized, for $(x_1,y_1),\ldots,(x_n,y_n)$ a sample of size n.

The parameter estimates provide a

EBWD(1.0133,8.091,7.2535,63.346,0.77468).

Table 1 includes the results of the χ^2-goodness of fit test (observed and expected frequencies), indicating the classes that have been grouped in order to consider expected values greater or equal than 5. The value of the χ^2-statistic (14.046) is less than the one obtained for the *BGWD* and, also, the p-value is higher (0.0806).

With regard to the components of the variance, the values obtained are included in Table 2. It should be noted that the majority of the variability is due to randomness. Moreover, the external factors or liability have less incidence than the internal factors or proneness in the explanation of the behavior of the number of accidents. It should be pointed out that even though the *BGWD* and the *EBWD* are different, the values obtained for the variance components are very similar to those obtained by Xekalaki, so it seems that both models coincide in the explanation of the factors that influence the number of accidents.

Table 1. Observed and expected values

	1931–33				
1934–36	0	1	2	3	4
0	23881 23887.9478	2117 2146.1793	242 214.6711	17 23.6536	2 0.4292
1	2386 2378.6215	419 418.1887	57 61.6481	9 8.9106	3 0.2410
2	275 260.5670	64 67.5159	12 13.0563	5 2.3224	1 0.0874
3	22 31.1481	5 10.5873	2 2.5195	2 0.5297	0 0.0264
4	5 4.0282	4 1.6850	0 0.4739	1 0.1145	0 0.0073

Table 2. Components of the variance

Components	1931–33	1934–36	1931–36
Randomness	0.1261 (86.3336 %)	0.1138 (87.2669 %)	0.2398 (78.5701 %)
Proneness	0.0160 (10.9505 %)	0.0130 (9.9876 %)	0.0579 (18.9580 %)
Liability	0.0040 (2.7159 %)	0.0036 (2.7457 %)	0.0075 (2.4719 %)
Total	0.1460	0.1304	0.3052

References

1. J.O. Irwing. (1968). The generalized waring distribution applied to accident theory. Journal of the Statistical Society, Series A, 131, 205.
2. A.W. Kemp and C.D. Kemp. (1975). Models for Gaussian hypergeometric distributions. Statistical Distributions in Scientific Work, 1, 31.
3. S. Kocherlakota and K. Kocherlakota. (1992). Bivariate Discrete Distributions. Marcel Dekker.
4. J. Rodríguez-Avi, A. Conde-Sánchez, M.J. Olmo-Jiménez and A.J. Sáez-Castillo. (2004). Properties and applications of the family of Gaussian discrete distributions. Proceedings of the International Conference on Distribution Theory, Order Statistics and Inference in Honour of Barry C. Arnold, Santander, Spain.

5. E. Xekalaki. (1983). The Univariate generalized waring distribution in relation to accident theory: Proneness, spells or contagion? Biometrics, 39, 887.
6. E. Xekalaki. (1984a). The Bivariate generalized waring distribution and its application to accident theory. Journal of the Royal Statistical Society, Series A, 147, 488.
7. E. Xekalaki. (1984b). Models leading to the Bivariate generalized waring distribution. Utilitas Mathematica, 25, 263.

Chapter 13

APPLYING A BAYESIAN HIERARCHICAL MODEL IN ACTUARIAL SCIENCE: INFERENCE AND RATEMAKING

J.M. PÉREZ-SÁNCHEZ

Department of Quantitative Methods in Economics
University of Granada, 18071-Granada, Spain

J.M. SARABIA-ALEGRÍA

Department of Economics, University of Cantabria, 39005-Santander, Spain

E. GÓMEZ-DÉNIZ

Department of Quantitative Methods in Economics
University of Las Palmas de Gran Canaria, 35017-Las Palmas de G.C. Spain

F.J. VÁZQUEZ-POLO

Department of Quantitative Methods in Economics
University of Las Palmas de Gran Canaria, 35017-Las Palmas de G.C. Spain

In a standard Bayesian model, a prior distribution is elicited for the structure parameter in order to obtain an estimate of this unknown parameter. The hierarchical model is a two way Bayesian one which incorporates a hyperprior distribution for some of the hyperparameters of the prior. In this way and under the Poisson-Gamma-Gamma model, a new distribution is obtained by computing the unconditional distribution of the random variable of interest. This distribution seems to provide a better fit to the data, given a policyholders' portfolio. Furthermore, Bayes premiums are thus obtained under a bonus-malus system and solve some of the problems of surcharges which appear in these systems when they are applied in a simple manner.

1. Introduction

From the Bayesian standard model point of view, a structure parameter follows a prior distribution. A hierarchical model is a two way Bayesian model which incorporates a hyperprior distribution for some of the hyperparameters of the prior. A new distribution is obtained by computing the unconditional distribution of the random variable of interest if the Poisson-Gamma-Gamma model is used. This distribution provides a better fit to the data. The hierarchical approach reflects a different statistical perspective on how to model the expert's information within the Bayesian framework. This Bayesian hierarchical methodology incorporates both the prior distribution and the data information into one unified modelling framework. In order to consider a hierarchical Bayes elicitation, we have to assume a framework in which structural and subjective prior information can be used to yield an elicited prior. In the hierarchical Bayes scenario, we have to specify our subjective beliefs about the hyperparameters of the prior distribution.

A Bayesian approach allows the statistician to compute the posterior probability for each model in a set of possible models. Using hierarchical approach, analysis can facilitate the choice of a satisfactory prior distribution. In this paper, we use this methodology in order to analyze its application to an insurance framework.

We apply the hierarchical model for computing bonus-malus premiums (BMP) in the same way as Lemaire [3]. Thus, hierarchical methodology incorporates knowledge about the number of claims believed a priori. The distribution of the number of car accidents in an automobile portfolio is known to be well fitted by a Poisson distribution, assuming that θ is the mean of the number of claims. Let us assume that the portfolio is not homogeneous and that the frequency of the risks is different in each case.

Bayesian hierarchical methodology is based on the use of hierarchical priors and we need to specify how the data (x) depends on the parameter of interest (θ), the likelihood function, $f(x \mid \theta, F)$, where x represents the sample information and F is an unknown parameter.

The prior specification is restricted to two stage priors:

• The standard prior distribution, $\pi_1(\theta \mid \lambda, G)$, where λ is a hyperparameter in Λ. This level indicates how the parameter of interest (θ) varies throughout the population, depending on two unknown parameters λ and G.

• The proper prior, $\pi_2(\lambda, F, G)$. In the second stage, instead of estimating θ, it will be considered as a random variable. In this level, we obtain a true prior density on the set of nuisance parameters, depending on λ, G and F.

The variables could be scalars, vectors or matrices, but here they are represented as scalars.

A third level distribution is to specify the posterior distribution of θ, or some features thereof. In this sense, we must specify the posterior distribution in terms of the posterior distributions at the various stages of the hierarchical structure. Therefore, we need to specify $\pi_2(\theta \mid x)$ in the third stage.

The main goal of a hierarchical Bayesian analysis is often to obtain the posterior distribution. If we apply the Bayes' theorem, we would obtain the posterior distribution in the following form:

$$\pi_2(\theta \mid x) = \frac{\iiint f(x \mid \theta, F)\pi_1(\theta \mid \lambda, G)\pi_2(\lambda, F, G)d\lambda dF dG}{\iiiint f(x \mid \theta, F)\pi_1(\theta \mid \lambda, G)\pi_2(\lambda, F, G)d\lambda dF dG d\theta}. \tag{1}$$

It is of great interest to estimate the posterior mean $E(\theta_i \mid x)$ and the variance $E(\theta_i^2)$. However, it is possible that the posterior distribution of λ, F and G is in our range of interest. In this case, we need to compute:

$$\pi_2(\mu, F, G \mid x) = \frac{\pi_2(\lambda, F, G)\int f(x \mid \theta, F)\pi_1(\theta \mid \lambda, G)d\theta}{\iiiint f(x \mid \theta, F)\pi_1(\theta \mid \lambda, F, G)\pi_2(\lambda, F, G)d\lambda dF dG d\theta}. \tag{2}$$

This model was introduced by Lindley and Smith [1]. More recently, Klugman [7] analyzed the normal-normal hierarchical structure from the Bayesian point of view. Cano [8] applied this methodology to study the Bayesian robustness of the model.

However, a continuous distribution is clearly inappropriate for frequency counts. For severe or total losses, the distribution places probability in negative numbers and so the Poisson and negative binomial are much more commonly used.

The rest of this paper is structured as follows: Section 2 analyzes a hierarchical Bayesian structure, the Poisson-Gamma-Gamma model. In Section 3 we use this model to compute premiums under a bonus-malus system. Section 4 applies the above results to an actuarial example. Finally, section 5 contains a discussion of related work.

2. Inference procedure

In this section, the hierarchical Bayesian Poisson-Gamma-Gamma model is studied. In this case, the hierarchical model is a two way Bayesian standard model which is built in the following way:

Firstly, we have the model depending on an unknown parameter θ, $f(x \mid \theta)$. We assume a Poisson distribution, i.e.,

$$f(x \mid \theta) \quad \sim \quad P(\theta). \tag{3}$$

Secondly, parameter θ follows a prior distribution which is assumed to be a Gamma distribution. Then:

$$\pi_1(\theta \mid a, b) \quad \sim \quad G(a, b), \quad a, b > 0 \tag{4}$$

where the Gamma distribution has a probability density function proportional to $\theta^{a-1}e^{-b\theta}$.

Thirdly, and finally, a gamma hyperprior distribution is assumed for the b parameter of the prior $\pi_1(\theta \mid a,b)$, i.e., $b \sim G(\alpha,\beta),\quad \alpha,\beta > 0$.

Therefore

$$\pi_2(b) \sim b^{\alpha-1}e^{-\beta\theta}. \tag{5}$$

It is well known that a mixture of distributions is a simple way to obtain new probability distributions. Thus, we can build the prior distribution of θ without depending on the b parameter, to obtain:

$$\begin{aligned}
\pi_1(\theta \mid a,\alpha,\beta) &= \int_0^\infty \pi_1(\theta \mid b)\pi_2(b)db \\
&= \int_0^\infty b^a\Gamma(a)\theta^{a-1}e^{-b\theta}\beta^\alpha\Gamma(\alpha)b^{\alpha-1}e^{-\beta b}db \\
&= \frac{1}{B(a,\alpha)\beta}\frac{(\theta/\beta)^{a-1}}{(1+\theta/\beta)^{a+\alpha}},
\end{aligned} \tag{6}$$

where $B(x,y)$ denotes the usual beta function.

This distribution corresponds to the Pearson type VI distribution, sometimes called second-kind beta distribution or beta-prime distribution, with scale parameter β (Stuart and Ord [5] and Johnson *et al.* [9]). A random variable with *pdf* (6) can be denoted by $\theta \sim B2(a,\alpha;\beta)$.

The moments of $\pi_1(\theta \mid a,\alpha,\beta)$ can be calculated by using:

$$\mu_r'(\theta) = \mathsf{E}(\theta^r) = \beta^r\frac{\Gamma(a+r)\Gamma(\alpha-r)}{\Gamma(a)\Gamma(\alpha)}, \qquad if\ \alpha > r \tag{7}$$

Thus, the mean and the variance are:

$$\begin{aligned}
\mathsf{E}(\theta) &= \frac{a\beta}{\alpha-1}, \qquad if\ \alpha > 1, \\
Var(\theta) &= \frac{a(a+\alpha-1)\beta^2}{(\alpha-1)^2(\alpha-2)}, \qquad if\ \alpha > 2.
\end{aligned}$$

These results are obtained under straightforward computations. An interesting property of the prior distribution in the Poisson-Gamma-Gamma model (or Poisson-second kind beta) is the over-dispersion it presents with respect to the classical Gamma distribution. In other words, when the mean of the Poisson-second kind beta distribution is equal to that of the Gamma distribution, the variance of the former is greater than that of the latter. This property gives the model more flexibility and makes it appropriate to use in a

BMS, where the variance of the observed data is generally greater than the mean (Shengwang *et al.* [12]).

The following proposition gives the posterior distribution of θ under the hierarchical Bayesian model.

Proposition 1 *The posterior distribution of* θ *given the data* x *in the hierarchical Poisson-Gamma-Gamma model is given by*

$$\pi(\theta \mid x) = \frac{e^{-t\theta}\left(\theta/\beta\right)^{a+x-1}\left(1+\theta/\beta\right)^{-(a+\alpha)}}{\Gamma(a+x)\mathcal{U}(a+x, x-\alpha+1, \beta t)}, \tag{8}$$

where

$$\mathcal{U}(m,n,z) = \frac{1}{\Gamma[m]}\int_0^{\infty} e^{-zs} s^{m-1}(1+s)^{n-m-1} ds, \qquad m, z > 0, \tag{9}$$

is the confluent hypergeometric function (Goovaerts and De Pril [4]).

Proof. It is straightforward by applying Bayes' Theorem.

3. Experience rating

To illustrate our approach, we apply the results obtained for computing premiums under a BMS. This is a merit rating method used in automobile insurance where the number of claims modifies the premium. A model often used for experience rating in a BMS assumes that each individual risk has its own Poisson distribution for a number of claims, assuming that the mean number of claims is distributed across individual policyholders (Coene and Doray, [10]; Corlier *et al.*, [2]; Lemaire, [3], [6], [11]).

A bonus-malus premium (BMP) can be computed under the variance principle (Gómez and Vázquez, [13]) in the same way as Lemaire [3] built a BMP under the net principle. In this sense, we have:

$$P_{BM}^{\pi_1,\pi_2}(x,t) = \frac{\int_\Lambda (\lambda+1)^2 \pi(\lambda \mid x) d\lambda}{\int_\Lambda (\lambda+1) \pi(\lambda \mid x) d\lambda} \frac{\int_\Lambda (\lambda+1) \pi(\lambda) d\lambda}{\int_\Lambda (\lambda+1)^2 \pi(\lambda) d\lambda} \tag{10}$$

Observe that this expression is simply a rate between a posterior magnitude and the corresponding prior. Next proposition gives the BMP in (10) under the model assumed in Section 2.

Proposition 2 Under the Poisson-Gamma-Gamma model the variance bonus-malus premium given in (10) is computed as:

$$P_{BM}^{\pi_1,\pi_2}(x,t) = K\frac{A+B+C}{D+C}, \tag{11}$$

where

$$\begin{aligned} A &= \beta^2(a+x+1)(a+x)\mathcal{U}(a+x+2, x-\alpha+1, \beta t), \\ B &= 2\beta(a+x)\mathcal{U}(a+x+1, x+\alpha+2, \beta t), \\ C &= \mathcal{U}(a+x, x-\alpha+1, \beta t), \\ K &= \left(\frac{\alpha\beta}{\alpha-1}+1\right)\left[\frac{a(a+\alpha-1)\beta^2}{(\alpha-1)^2(\alpha-2)} - \frac{a^2\beta^2}{(\alpha-1)^2}\right]^{-1}, \end{aligned}$$

and $\mathcal{U}(m,n,z)$ is the confluent hypergeometric function defined in (9).

Proof. It is straightforward to prove this proposition by:

$$\int_\Lambda (\lambda+1)\pi(\lambda\,|\,x)d\lambda = \frac{\beta(a+x)\mathcal{U}(x+a+1, x+\alpha+2, \beta t)}{\mathcal{U}(a+x, x-\alpha+1, \beta t)} + 1,$$

and

$$\begin{aligned} \int_\Lambda (\lambda+1)^2\pi(\lambda\,|\,x)d\lambda &= \int_\Lambda \lambda^2\pi(\lambda\,|\,x)d\lambda + 2\int_\Lambda \lambda\pi(\lambda\,|\,x)d\lambda + 1 \\ &= \beta^2(a+x+1)(a+x)\frac{\mathcal{U}(x+a+2, x-\alpha+1, \beta t)}{\mathcal{U}(x+a, x-\alpha+1, \beta t)} \\ &\quad + 2\beta(a+x)\frac{\mathcal{U}(x+a+1, x-\alpha+2, \beta t)}{\mathcal{U}(x+a, x-\alpha+1, \beta t)} + 1. \end{aligned}$$

Although we do not have a perfect closed form for this BMP, its computation is simple by using, for example, MATHEMATICA software, because the confluent hypergeometric function is tabulated.

4. Numerical example

In this section, the results obtained in the preceding sections are illustrated with an example from Lemaire [3], which represents the claims made by policyholders of a Belgium insurance company during four periods.

Figure 1 shows the distribution for the number of claims, which provides a fairly good fit, accepted by the χ^2-test of goodness of fit.

The mean and variance of this distribution are 0.1011 and 0.1074, respectively. The parameters of the structure function were estimated by applying the method of moments. The estimated parameters are $\hat{a} = 3.25585$, $\hat{\alpha} = 6.13732$ and $\hat{\beta} = 0.159492$.

The results are illustrated in Table 1, which shows the BMP for the hierarchical structure considered (in bold) and the BMP for the standard Bayesian methodology.

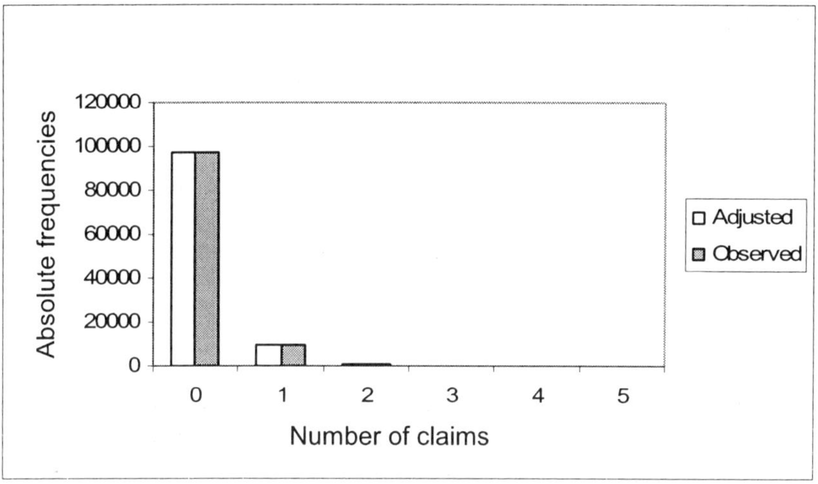

Figure 1. Observed distributions

Table 1. Bonus-malus premiums under both standard and hierarchical models

	x			
t	0	1	2	3
1	0.994	1.050	1.105	1.161
	0.993	**1.048**	**1.131**	**1.265**
2	0.998	1.041	1.094	1.146
	0.988	**1.036**	**1.104**	**1.202**
3	0.984	1.033	1.083	1.133
	0.984	**1.027**	**1.086**	**1.164**

It is clear from Table 1 that the relative premiums allow the transition rules commented above. For example, a policyholder has to pay 1.104 monetary units in the second period because of his/her two previous claims. In the next period, the policyholder will have to pay 1.164 monetary units if he/she makes a claim. However, the premium will decrease to 1.086 monetary units if he/she does not make a claim. This behaviour is observed for all the premiums, and so we obtain BMP by using a hierarchical Bayesian model.

Table 2 shows how a hierarchical BMP gives a bonus to good drivers with respect to standard Bayesian premiums by decreasing their percentage of penalization for the transition $x = 0 \to x = 1$ and $t = 1 \to t = 2$. However, the hierarchical structure increases the percentage of penalization for the other transitions.

Table 2. Percentage of penalization

	Δx		
$1 \to 2$	4.7%	10%	15.2%
	4.3%	**11.1%**	**20.9%**
$2 \to 3$	3.5%	8.9%	13.9%
	3.9%	**9.3%**	**17.1%**

5. Conclusions

In this article we review some aspects of the hierarchical Bayesian models and emphasize the Poisson-Gamma-Gamma model because of its practical use in actuarial science. In order to model the number of claims of a BMS, we use a hierarchical structure in which the second-kind beta distribution arises as the hyperprior distribution.

The model poses no additional complications, as many of its positive properties can be deduced analytically. The model can be applied straightforwardly to actuarial premium-setting problems, and we show that these premiums follow the transition rules of BMS. These transition rules allow the *malus* policyholders to be surcharged and a bonus given to the *bonus* ones.

In order to check the prior distribution, we can carry out a Bayesian robustness analysis of the premiums in the same way as Gómez and Vázquez [13]. These authors studied the sensitivity of a BMS from a standard Bayesian point of view. In the hierarchical setting, a Bayesian robustness analysis can be carried out in the same way as in Cano [8], where the normal-normal hierarchical model is analyzed.

References

1. D.V. Lindley and F.M. Smith. (1972). Bayes estimates for the linear model. Journal of the Royal Statistical Society B, 34, 1–41.
2. F. Corlier, J. Lemaire and D. Muhokolo. (1979). Simulation of an Automobile Portfolio. Essays in the Economic Theory of Risk and Insurance, 11, 40–46.

3. J. Lemaire. (1979). How to define a bonus-malus system with an exponential utility function. Astin Bulletin, 10, 274–282.
4. M.J. Goovaerts and N. De Pril. (1980). Survival probabilities based on Pareto claim distributions. Astin Bulletin, 11, 154–157.
5. A. Stuart and J.K. Ord. (1987). Kendall's Advanced Theory of Statistics (Vol. 1, Chapter 6). New York: Oxford University Press.
6. J. Lemaire. (1988). Construction of the new Belgian motor third party tariff structure. Astin Bulletin, 18(1), 99–112.
7. S. Klugman. (1992). Loss Model from Data to Decisions. New York: Willey.
8. J.A. Cano. (1993). Robustness of the posterior mean in normal hierarchical models. Communications in Statistics, 22(7), 1999–2014.
9. N.L. Johnson, S. Kotz and N. Balakrishnan. (1995). Continuous Univariate Distributions (vol. 2, second edition, chapter 27). John Wiley, New York.
10. G. Coene and L. Doray. (1996). A financially balanced Bonus-Malus system. Astin Bulletin, 26, 107–116.
11. J. Lemaire. (1998). Bonus-Malus system: The European and Asian approach to merit-rating. (With discussion by Krupa Subramanian, "Bonus-Malus system in a competitive environment"), North American Actuarial Journal, 2(1), 1–22.
12. M. Shengwang, W. Yuan and G. Whitmore. (1999). Accounting for individual over-dispersion in a bonus-malus automobile insurance system. Astin Bulletin, 29(2), 327–337.
13. E. Gómez and F.J. Vázquez. (2005). Modelling uncertainty in insurance bonus-malus premiums principles by using a Bayesian robustness approach. Journal of Applied Statistics, 32(7), 771–784.

Chapter 14

ANALYSIS OF THE EMPIRICAL DISTRIBUTION OF THE RESIDUALS DERIVED FROM FITTING THE HELIGMAN AND POLLARD CURVE TO MORTALITY DATA

F. ABAD-MONTES

Dpto. Estadística e Investigación Operativa, Universidad de Granada
C/Fuentenueva, s/n, Granada, España

M.D. HUETE-MORALES

Dpto. Estadística e Investigación Operativa, Universidad de Granada
C/Fuentenueva, s/n, Granada, España

M. VARGAS-JIMÉNEZ

Dpto. Estadística e Investigación Operativa, Universidad de Granada
C/Fuentenueva, s/n, Granada, España

In studying the behaviour of human phenomena, it is of interest to examine the patterns that remain more or less stable, whether the comparison is made of different populations at a given moment or at different times, or of the same population in different situations. Such regularities have long been modelled, and this has enabled researchers to discover aspects and properties that are inherent to the phenomenon being studied. In the present paper, various techniques, some of which are relatively modern, are applied to the analysis of the empirical distribution of the residuals derived from fitting the Heligman and Pollard curve to mortality data. Firstly, we perform a graphical illustration from the time perspective (curves fitted over various periods) and then a static one for the ages (i.e. obtaining fits to different ages). The aim of this study is to explore the different distributions of the residuals at each age and thus to evaluate the correspondence between models (such as the Heligman and Pollard curve) and reality (the observed rates of mortality). For this purpose, we use graphical techniques, non-parametric techniques such as kernel smoothing, splines and weighted local fit, and generalised additive models, together with bootstrap sampling techniques to describe distributions of statistical measures of the residuals.

1. Introduction

It is frequently necessary to determine the density function of certain data sets, especially when such data present characteristics which, a priori, cannot be assumed to behave like standard probability models. The experience of demographers in fitting the Heligman and Pollard (H-P) curve to rates of mortality shows that there exists a biased systematic pattern in the values fitted for given age ranges. This curve provides a fairly good description of the behaviour of mortality rates as a function of age, and thus it is widely used.

However, in the present study we seek to better identify the limitations of forecasts made using H-P fitting, by means of a statistical analysis of the behaviour of the distribution of residuals.

Analysis of such Heligman and Pollard residuals (rHP) was carried out using standard current techniques to estimate the properties of distributions from a perspective that is basically non parametric.

2. Data

We took the rHP residuals derived from the results of fitting H-P curves to the mortality rates, qx, observed for ages 0 to 84 years for the population of Andalusia for the period 1976–2002.

3. Exploration and graphical summary of the distributions of the residuals

3.1. *Behaviour of the residuals in each period (H-P fit)*

Apart from a few anomalous points, the average behaviour is analogous in each fit. The curves are assumed to be fitted in a similar way in each period observed.

The fit of a spline shows a line close to the zero line, in accordance with the previous figure.

In short, the above figures show a similar behaviour pattern for the residuals derived from an H-P fit for each curve fitted for the corresponding period.

3.2. *Behaviour of the fits for each age group*

The figure shows the differences between the distributions of the residuals for each age group: Box graph and Dispersion or scatter plot diagram

The same cannot be said of the pattern of the residuals when we consider the pattern of the distribution of each age that is examined.

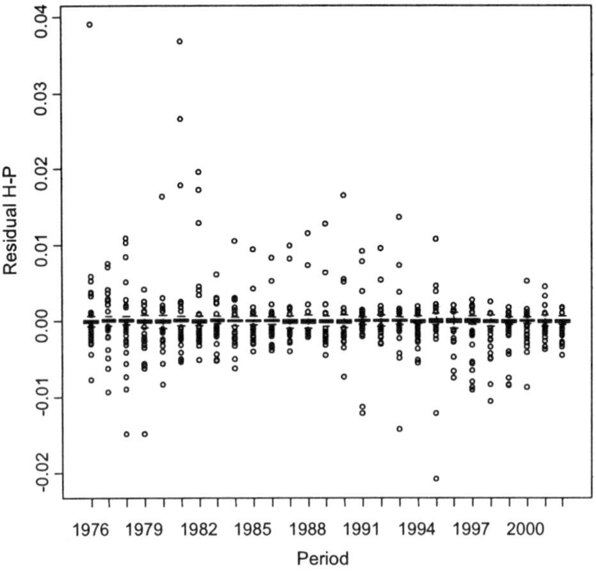

Figure 1. Distribution of the H-P residuals by period

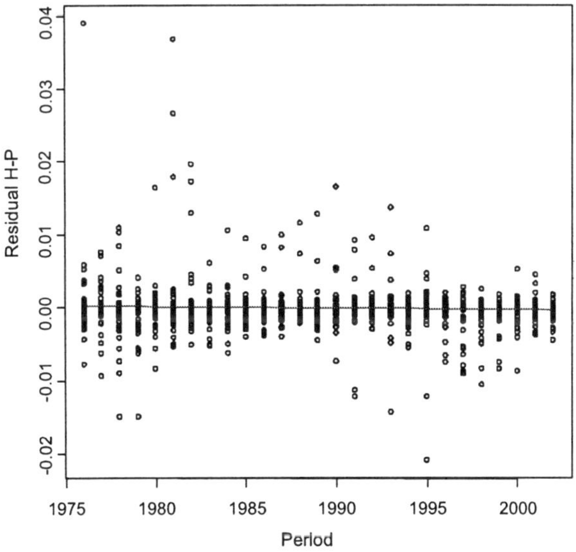

Figure 2. Distribution of the H-P residuals over the period

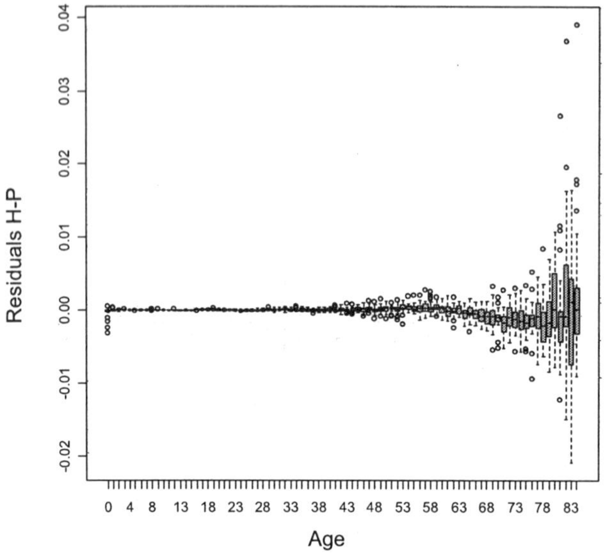

Figure 3. Distribution of the residuals for each age

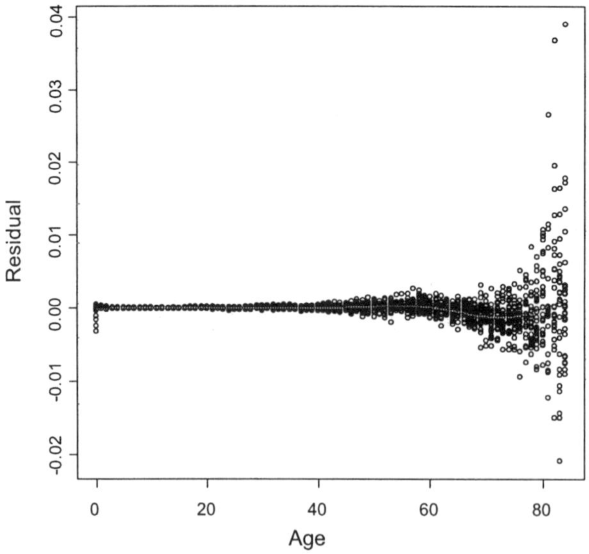

Figure 4. Distribution of the residuals by age

3.3. *Behaviour pattern of means and variances of the residuals according to the age and period examined*

The next figure shows the systematic behaviour pattern of the means and variances of the residuals according to the age at which the fit to the mortality rate is carried out. The trend of the latter is seen to be less regular for the fit in relation to the period.

to the age at which the fit to the mortality rate is carried out. The trend of the latter is seen to be less regular for the fit in relation to the period.

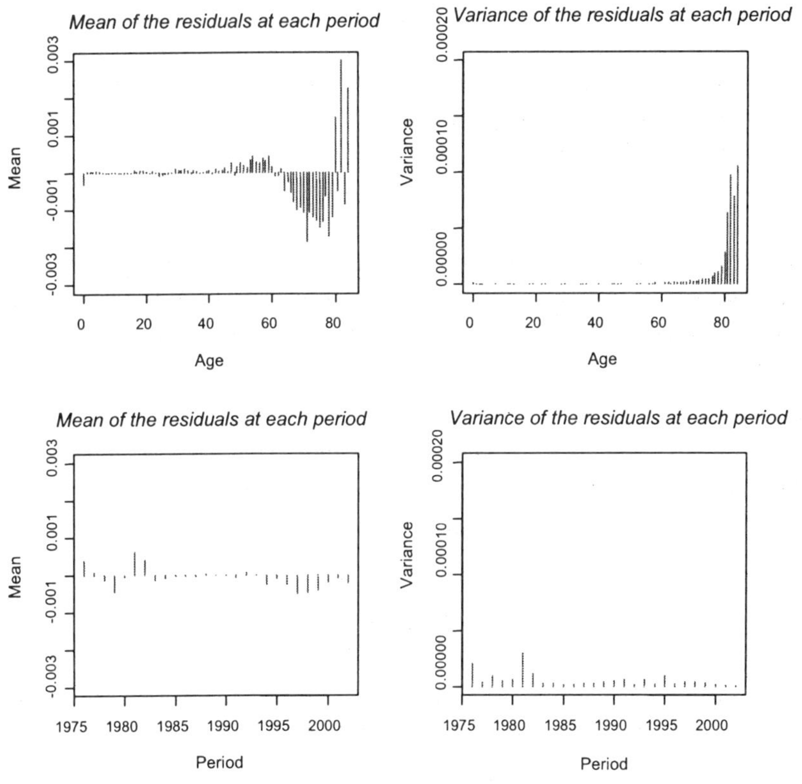

Figure 5. Means and variances for ages and periods

The top left figure shows that the assumption that the distribution of the residuals presents an approximately zero mean for each age is unlikely to be fulfilled.

It can be seen that the curves are not fitted in the same way for every age; at some (60–80 years), the figures show the residuals to be systematically negative. Another noteworthy aspect is the diversity in the variability.

4. Non-parametric regression curves

Sometimes it is impossible to model a function using parametric techniques. The *nube de puntos* diagram showing residuals versus age seems to show non-linear effects of age versus the value of the residuals. There are, however, flexible methods of describing such non-linear relationships, namely non-parametric regression techniques. Different algorithms for fitting non-parametric curves enable us to represent the effects of independent variables without specifying the global shape of the relationship, which facilitates a clearer visual interpretation of local behaviour patterns.

Assume a sample (x_i, y_i) i=1, …, n of values in the variables X and Y.

Let us denote the relation between x and y by

$$Y = \mu(x) + \varepsilon \tag{1}$$

where $\mu(x)$ describes an unknown function of x, representing the trend underlying the data, which is normally a smoothed trend fitted to the *nube de puntos*, and which can be estimated by various methods, among which the following are the most widely used:

a) The results obtained from locally averaging the response values observed in a range of values close to x, as **kernel smoothing**. This method produces an estimation of the mean response of Y in x, by means of the following ratio:

$$\hat{\mu}(x) = \frac{\sum_i y_i k\left(\frac{x - x_i}{b}\right)}{\sum_i k\left(\frac{x - x_i}{b}\right)} \tag{2}$$

where the nucleus function k is a function of the symmetric density (normally, the standardised normal) and where b is a constant that determines the size of the averaging operation; its value represents a midway position between an estimation that is more or less biased and which contains a greater or lesser degree of variability. The weights used for calculating the average of the response values decrease with increasing distance from point x.

b) The results obtained from the weighted local fit of a *p*-type polynomial. Several variations of this method have been developed, including **loess** and **locpoly**, implemented in R, which differ from each other in the parameter used for the smoothing process. Cleveland's local regression method (loess) establishes a neighborhood at point x, determining a proportion of points (the span) to be used to estimate the mean response of Y at this point.

The **loess** function enables us to achieve a local fit that adapts more flexibly to the trend of the data. It occupies an intermediate position between the global fitting of a function (linear, quadratic, cubic, etc.) and local fitting based on averaging the points (the calculated percentage of the total n) which constitute the closest neighborhood to each fitted point. Higher span values correspond to smoother curves.

The method consists, for each point x to be fitted, of performing a **weighted** regression of a curve, whether linear or polynomial, to the proportion of points closest to x thatcomprise the neighborhood in question. The weights reflect the proximity to or distance from the point with the tri-cube function , and are assigned to each point of the neighborhood.

Given a point x_i of the neighborhood of x, let $M(x) = m\acute{a}x|x - x_i|$ be the maximum distance for the points x_i of the neighborhood of x.

The weight of each point of the neighborhood is equal to

$$w(x) = \left(1 - \left(\frac{|x - x_i|}{M(x)}\right)^3\right)^3 \tag{3}$$

The method implemented in R, **locpoly**, enables both a regression fit and a density function to be obtained. In the process of fitting the local polynomial, it utilises the kernel weights derived from a k function (normally the standardised normal)

$$k\left[\frac{x_i - x}{b}\right] \tag{4}$$

the values of which decrease as x_i becomes more distant from x.

The value of the estimated curve is equal to the intercept of the fitted local polynomial, and is obtained by minimising the sum of the weighted squares.

$$\sum_{i=1}^{n}\left[y_i - (\beta_0 + \beta_1(x_i - x) + \beta_2(x_i - x)^2 + \ldots + \beta_p(x_i - x)^p)\right]^2 k\left[\frac{(x_i - x)}{b}\right] \tag{5}$$

Assuming that the weights matrix at a point x is

$$W(x) = Diag\left\{k\left[\frac{(x_i - x)}{b}\right]\right\} \tag{6}$$

and that the matrix X evaluated at the point x is

In addition to the base functions of truncated polynomials, others can be used. Indeed, in practice, it tends to be more useful and easier to implement certain bases which produce equivalent results. One of the main disadvantages of truncated polynomial bases is their lack of orthogonality and the instability that can arise when many knots are used.

Among the possible base functions are B-splines, which are useful because of their property of numerical stability, in comparison with the base of series of truncated powers and natural splines, which are linear in the intervals $(-\infty, a_1]$, $[a_K, \infty)$, where a_1 and a_K are the first and last (external) knots.

The spline functions implemented in the statistical software R, bs() and ns(), generate the bases of B-splines and of natural splines, respectively, that can be used in regression.

By means of the **smoothing spline** method, a compromise can be obtained between the degree of fit of the curve to the data and the smoothing of the shape. It is not constructed explicitly, but rather is obtained as the solution to an optimisation problem.

It is estimated using the criterion of penalised least squares, which minimises the sum of the squared residuals and penalises them using the integral of the squared second derivate, thus taking into account the degree of curvature of the estimated function.

$$\sum_i [y_i - \mu(x_i)]^2 + \lambda \int [\mu''(x)]^2 dx \qquad (16)$$

It has been shown that the minimiser is a cubic spline with knots at the n points x_i.

A cubic spline with knots at the n points x_i of the sample does not interpolate the data if lambda is greater than zero. The greater or lesser value of lambda controls the smoothing of the curve. A small lambda value gives rise to a curve that fits closely or interpolates the sampled data, while a large value produces a parametric fit that is dependent on the baseline functions of the spline.

The goodness of the fit will depend on the degree of the polynomial elements, on the number of knots and on the value of the lambda parameter that is used for smoothing. This lambda value has a great influence on the results of the fitting. By varying the lambda value, from lesser to greater, we can see, on a two-dimensional figure, how the curve tracks a trend that is perhaps clearer but at the cost of being less well adapted to the whole data set. The choice of the most appropriate lambda value is a difficult one. There are automatic procedures for this, which are based on the nature of the data, and one of the most commonly used such procedures is that of cross validation (CV).

The cross validation technique consists of dividing the data set into two parts: one that is used to estimate the model and another that enables us to make a prediction. Thus, the values that are used for predicting do not play any part in the fitting procedure. A particular case consists of reserving a single observation for predicting, the rest (n-1) being used to estimate the model, in each of the n partitions created.

Given n values in the response Y: $y_1, \ldots, y_n$ and the corresponding predicted values, $\hat{y}_{-1},\ldots,\hat{y}_{-i}\ldots\hat{y}_{-n}$, CV is defined as the sum of the squared residuals:

$$CV = \sum_{i=1}^{n}(y_i - \hat{y}_{-i})^2 \tag{17}$$

where $\hat{y}_{-i}$ is the predicted value of the i-th case, when this case has not been used to estimate the model.

In particular, given a lambda value and the predicted value of x_i in the non-parametric regression curve, computed without the observation (x_i, y_i), which we shall denote as $\hat{\mu}_{\lambda,-i}\ (x_i)$, then the following definition may be made:

$$CV_\lambda = \sum_{i=1}^{n}(y_i - \hat{\mu}_{\lambda,-i}\ (x_i))^2 \tag{18}$$

What is chosen is the lambda value that minimises CV.

In most statistical programs that implement this procedure, the fit is obtained by specifying the degrees of freedom of the curve or by applying cross validation.

The splines described above can be presented in the form

$$\hat{\mu} = S_\lambda y \tag{19}$$

They are described as linear because they are linear functions of the data vector **y**, where the matrix S does not depend on y.

The lambda parameter is difficult to interpret, but a transformation of this value, given by the trace of the matrix S, also reflects the amount of smoothing applied to the curve. Under standard (parametric) regression analysis, the trace of matrix H (matrix hat) is equal to the number of parameters fitted, which is equal to the degrees of freedom of the fit. In a similar way, the trace of S can be seen as a generalisation of this concept, being interpreted as "equivalent" degrees of freedom of the fit.

5. Estimating the probability density function

Estimation of the density using the kernel method is done by means of the expression:

$$\hat{f}(x) = \frac{1}{nh}\sum_{i=1}^{n} k\left(\frac{x - x_i}{h}\right) \tag{20}$$

estimating f(x) for a random sample $x_1, \ldots, x_n$, where k is a symmetric density function, for example, the standardised normal function. The value h is usually large enough so that excessive smoothing is not produced, thus avoiding the elimination of significant modes, but not so small as to allow too many random spikes. A large value would lead to an excessively biased estimate, while a low one would produce an estimate with too much variability. The choice of h is not immediate. Some authors have proposed the execution of various solutions in order to determine an optimum value. The method implemented in R is proposed by Sheather and Jones (1991).

The following figures show an initial approximation of the density function.

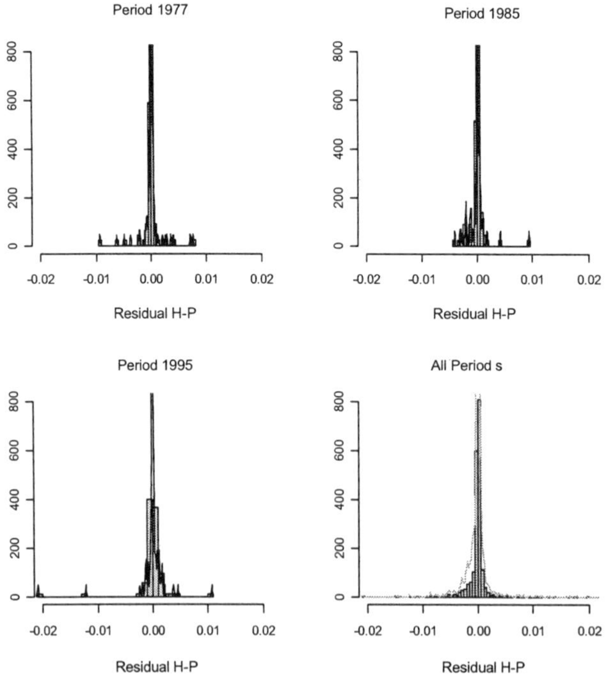

Figure 6. Distributions of residuals by periods

Although the sample size is small, we can see the high degree of similarity in the pattern of the probability density function in each period, with similar ranges of variability and similar function shapes.

A graphic examination of the distribution, according to the age of the subject, reveals patterns that are much more varied.

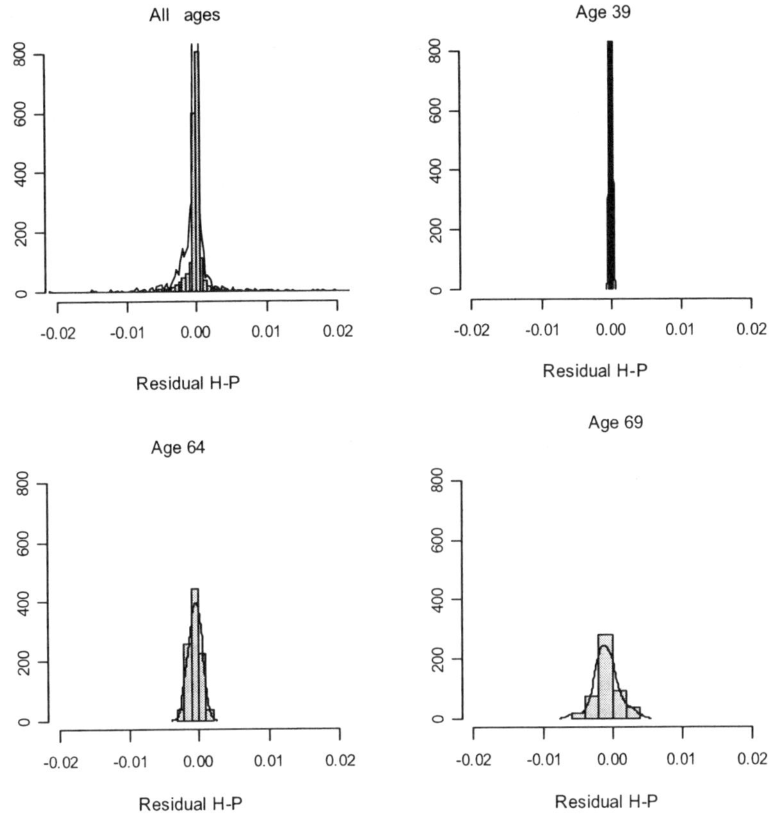

Figure 7. Distributions of residuals by ages

The above figure shows various shapes and differing ranges of variability in the density functions that were estimated for different age values.

6. Statistical inference based on the empirical distribution

It is of interest to estimate aspects of the probability distribution F of the residuals, based on a sample of size n. The estimated empirical distribution of F

For i=1,...,n, the jackknife i-th sample, denoted by x(-i), is obtained by eliminating the i-th element x_i, from the observed sample. The i-th replica of the statistic t(x) is based on this sample and is the partial estimator or statistic that is evaluated in this sample, t(x(-i)). Therefore it uses the empirical distribution of the n-1 points in x(i). Thus, we obtain the following set of pseudovalues that represent a new sample:

t*(x(1)), t*(x(2)), ...,t*(x(n)), where t*(x(i))= n t(x) - (n-1) t(x(-i)) for i=1,...,n

The jackknife estimator of Θ is obtained as the mean of these pseudovalues:

$$t^*(x) = \frac{\sum_{i=1}^{n} t^*(x(i))}{n} \tag{24}$$

The variance of the jackknife estimator is obtained in a similar way to that used to derive the variance of a sample mean.

$$\frac{\sum_{i=1}^{n} \left[t^*(x(i)) - t^*(x)\right]^2}{n^2} \tag{25}$$

Jackknife-influence values are established for the n values of the sample at differences of t(x(-i)) - t(x) for i=1, ..., n.

The techniques known as **"jackknife-after-bootstrap"** consist of applying jackknife to the results generated by the bootstrap method. One means of checking or diagnosing the degree of influence of a given observation x_i of the sample on the value of the statistic t used in bootstrap is the **jackknife-after-bootstrap figure**. This method enables us to detect the changes produced in the empirical quantiles of t*-t if an observation x_i is eliminated from the sample. Specifically, we construct a figure with various quantiles (such as 0.05, 0.10, 0.16, 0.5, 0.84, 0.9, 0.95) that are determined using bootstrap with all the values of the original sample and represented by horizontal lines. Each of the n x_i points of the sample is represented with abscissas that are equal to the corresponding values of empirical influence (for example, the jackknife value obtained by regression) and with ordinates that are equal to the value of the difference between the quantile obtained with the complete bootstrap simulation and the quantile obtained with simulations from which x_i is absent[1].

Note: The influence function or influence component can be considered a type of derivate that reflects the change in t(F) when the distribution F is subjected to a small contamination in x. These values are useful for determining the approximate variance of a statistic taking into account that such a statistic may be a kind of first order expansion of a Taylor series (for more information, see Efron and Tibshirani, 1993, pp. 298–302).

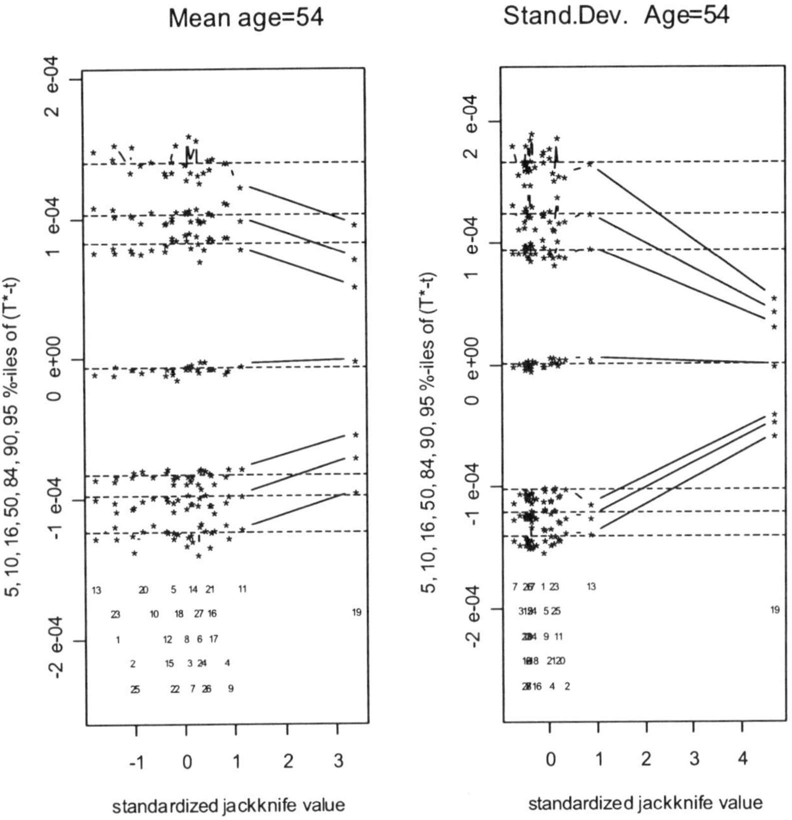

Figure 12. Jacknife-after-bootstrap (mean and stand. dev. age =54)

The figure highlights the noticeable effect of Observation 19. The sensitivity of bootstrap techniques to anomalous values makes it advisable for these to be eliminated.

7. Aspects of inference in non-parametric regression

7.1. *Confidence intervals in splines*

As in parametric regression, questions may be raised concerning inference on the fitted curve. Specifically, we are interested in obtaining confidence intervals for values fitted on the curve.

Confidence and prediction intervals for fitted values

Given the model

$$y = \mu + \varepsilon \tag{26}$$

and assuming that for a given smoothing parameter, the non-parametric regression curve can be stated in the linear form

$$\hat{\mu} = Sy \tag{27}$$

The covariance matrix of the fitted vector, $\hat{\mu}$, is $Cov(\hat{\mu}) = SS'\sigma^2$. Given an estimator of the residual variance, $\hat{\sigma}^2$, we can obtain an estimation of this matrix, the diagonal elements of which represent the estimated variances of the components of the vector $\hat{\mu}$, by merely replacing σ^2 with $\hat{\sigma}^2$ in the expression of the above covariance matrix. Thus it is possible to derive confidence intervals in a similar way to parametric regression, as well as prediction intervals for new estimations of the dependent variable.

If the errors ε in the model $y = \mu(x) + \varepsilon$ are normal and with a constant variance σ^2, the intervals are defined by:

$$\hat{\mu}(x_0) \pm z_{1-\alpha/2}\hat{\sigma}_{\hat{\mu}(x0)} \tag{28}$$

where $\hat{\sigma}_{\hat{\mu}(x0)}$ is the standard deviation for the value fitted in x0, $\hat{\mu}(x_0)$, obtained from the square root of the estimated variance, $\hat{V}(\hat{\mu}(x_0)) = S'_{x0}S_{x0}\hat{\sigma}^2$, where the row vector of S, termed S_{x0}, defines the linear combination of values of **y** such that $\hat{\mu}(x_0) = S_{x0}'y$.

For a small sample size, the Student t may be replaced by the normal value. The degrees of freedom are those appropriate for the closest integer, and correspond to the residual part of the fitted model. If the errors are not normal and if n is large enough, the intervals given above may continue to be valid, because of the central limit theorem.

The **prediction intervals** are also derived in a similar way to the parametric regression, that is, by means of

$$\hat{\mu}(x_0) \pm z_{1-\alpha/2}\hat{\sigma}\sqrt{1+S'_{x0}\,S_{x0}} \tag{29}$$

Logically, these are broader because they reflect, additionally, the uncertainty in the observation about its mean.

The estimated value of the residual variance σ^2 is obtained in a similar way to the parametric regression, as the ratio of the sum of the squares of the residuals (SSR) and the associated degrees of freedom. In parametric regression we find that the expected value of the SSR is equal to the product (n-p) σ^2, where p is the number of parameters in the model. Thus, we obtain as the estimator of the variance of the residuals the ratio

$$\frac{SSR}{n-p} \tag{30}$$

In non-parametric regression, it can be shown that the expectation of the sum of the squares of the residuals is approximately

$$E(SSR) \approx \sigma^2\left[track(SS') - 2track(S) + n\right] \tag{31}$$

and we obtain as the estimation of the variance of the residuals the following ratio:

$$\hat{\sigma}^2 = \frac{SSR}{g.l._{resid}} = \frac{SSR}{n - 2track(S) + track(SS')} \tag{32}$$

In fact, the intervals that are constructed in this way are not really confidence intervals for $\mu(x_0)$, but are for the expected value of $\hat{\mu}(x_0)$. They can only be interpreted as confidence intervals for $\mu(x_0)$ if there is no inherent **bias** in the regression curve, and this is very difficult to detect. Therefore, it is more appropriate to use the term **bands of variability**, and a value of 2 is normally used for z. In practice, prediction intervals are usually interpreted as confidence intervals, because the bias is usually small compared to the variability, and therefore can be ignored.

It should be remembered that these intervals cannot be interpreted as descriptors of the global characteristics of the entire curve, as they only reflect the behaviour pattern of each point that is estimated.

The following figure shows the variability bands of the fitted curve:

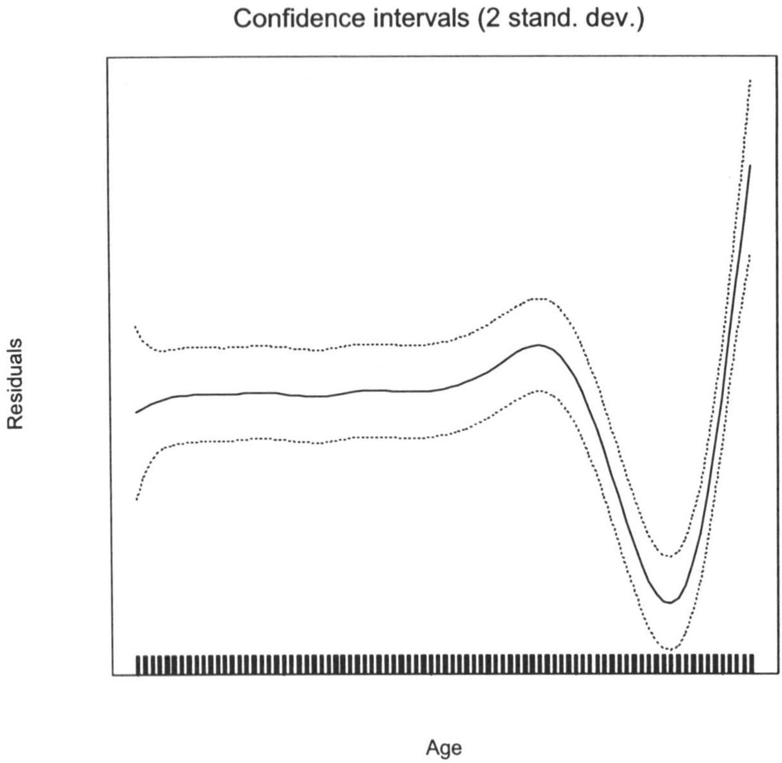

Figure 13. Variability bands of de fitted curve

7.2. *Confidence intervals (bootstrap procedure)*

As remarked above, one of the problems inherent in fitting a non-parametric curve is that of bias. For the above-constructed intervals to be interpreted as confidence intervals it is necessary for the fitted curve to be free of bias.

In constructing confidence intervals, various authors have proposed more or less complex methods to take the above consideration into account. A strategy is also employed in bootstrap methods, for example that of using a smoothing parameter that is small enough to simulate the residuals of the fit and thus reduce the degree of bias, but describing a curve with a shape that is less smoothed than that corresponding to an optimum smoothing parameter derived from a cross-validation criterion.

In a similar way to the use of bootstrap methods in standard regression, here we carry out a bootstrap simulation of splines fitted to the residuals as a function of age, in order to determine the mean expected residual for a given age.

To do so, we take 999 bootstrap samples, in order to obtain confidence intervals for the values fitted to given ages. Specifically, once we have fitted a spline to the data, and in order to take into account the problem of bias that is inherent to these non-parametric regression methods, the following simulation mode is now adopted.

From the optimum smooth parameter derived from the fit to the data of the original sample, we obtain a spline that produces a greater degree of smoothing in the model, using a duplicate of the original parameter, the estimations of which, therefore, present less variability. Moreover, we determine another spline that produces greater variability in the estimations, and therefore reduces the bias.

The residuals derived from these latter splines (with greater variability), sampled with replacement, are used to generate the new sets of responses, being added to the values fitted with the spline obtained with the duplicated smooth parameter.

The following figure shows the confidence intervals resulting from the simulation using this strategy, which enables us to alleviate the problem of bias.

The intervals were determined for certain ages. Note that in some of these the interval does not contain the value zero.

Age: 37, 54, 75
Fitted value: 3.427417e-05, 3.125760e-04, -1.132426e-03
Lower limit: -0.0001572603, 0.0001310923, -0.0013644013
Upper limit: 0.0001851737, 0.0004741195, -0.0009346646
Confidence level: 90%

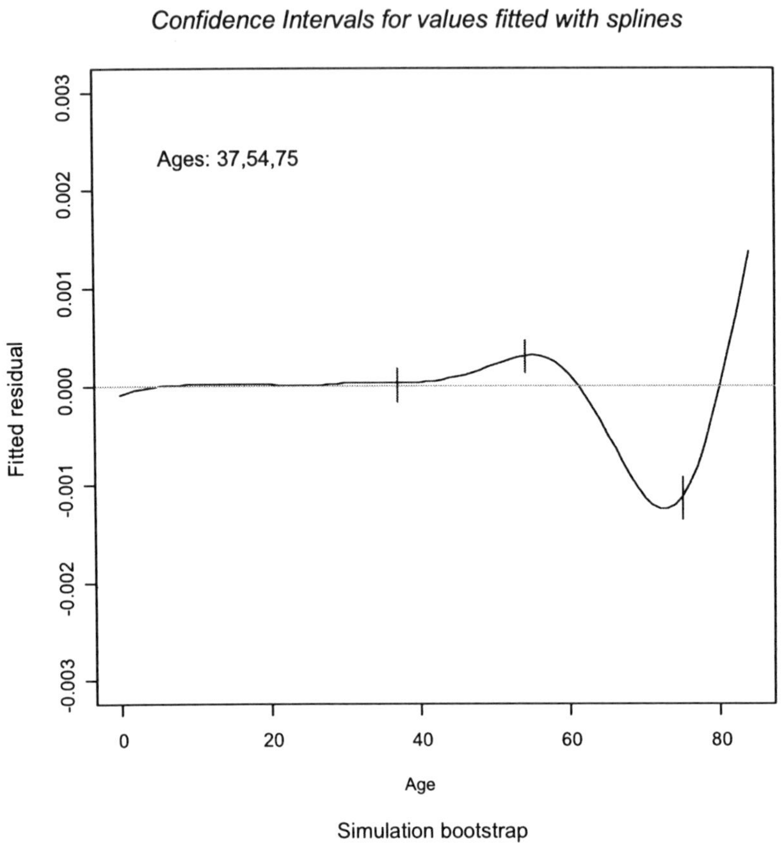

Figure 14. Confidence intervals for values fitted with splines

7.3. *Comparison of linear and non-parametric models*

It can be seen, moreover, that the proposed model with a non-parametric curve produces a better fit than does the linear model. This contrast of the models is highly significant.

The following results show that the replacement of a slope by the smooth curve gives rise to a significant reduction in the residual part of the model, which demonstrates that the residuals present a non-linear dependence with age, which corroborates the above graphic studies.

The contrast statistic used approximates an F distribution. Thus, given the following models:

Model 1: linear model that expresses residuals as a function of age;

Model 2: smooth spline

The statistic is given by the ratio

$$\frac{(SSR1 - SSR2)/(g.l.2 - g.l.1)}{SSR2/(n - g.l.2)} \rightarrow F_{g.l.2-g.l.1,\, n-g.l.2} \tag{33}$$

where SSR1 and SSR2 are the sums of the squares of the residuals in Models 1 and 2, respectively; g.1.1 and g.1.2 are the corresponding degrees of freedom. Thus, we obtain a significance of the order of 2.2e-16.

8. Brief review of generalised linear models, additive models and generalised additive models

Generalised linear models

Generalised linear models (GLM) are an extension of linear models, and their characteristics enable the application of a unified statistical approach, based on the common structure they share.

The linear model of normal errors $Y = X\beta + \varepsilon$ is extended to responses with other distributions within the exponential family, which also includes the normal distribution, to facilitate the modelling of variables with densities belonging to this family, whether discrete variables such as Poisson and binomial ones or continuous ones such as the normal and gamma types.

The GLM model consists of a random component, Y, a systematic one or linear predictor $\eta = X\beta$, and a link function, g, to relate them.

The distribution of Y has a density function that takes the following shape:

$$f_Y(y_i) = \exp\left\{\frac{y_i\theta_i - b(\theta_i)}{\Phi} + c(y_i, \Phi)\right\} \tag{34}$$

θ is called a natural parameter, while Φ is the dispersion parameter.

The link function, g, relates the mean of Y, $\mu = E(Y)$, with the linear predictor $\eta = X\beta$ by means of $g(\mu) = X\beta$.

The solution to the parameters, β, of the model is obtained by maximum likelihood estimation. The solution derived from applying weighted least squares to a new response variable, Z, is given by:

$$\beta = \left[X' Diag\left\{ \frac{1}{Var(z_i)} \right\} X \right]^{-1} X' \left[Diag\left\{ \frac{1}{Var(z_i)} \right\} \right]^{-1} Z \quad (35)$$

where the dependent variable Z is:

$$Z = X\beta + Diag\{g'(\mu_i)\}(y - \mu) \quad (36)$$

(this can be considered a first-order approximation to a Taylor series), and the variance is expressed as:

$$Var(Z) = \Phi Diag\{v(\mu_i)[g'(\mu_i)]^2\} \quad (37)$$

with the weights being determined from the inverse values of the variance.

Additive models

In an additive model, the linear terms $\beta_j X_j$ in the expression representing the linear predictor $X\beta$ of the linear model, are replaced by functional terms f_j that are smoothed and non linear. In this sense, the additive model can be considered an extension of the linear model. It is described by the following expression:

$$Y = \alpha + f_1(X_1) + ... + f_p(X_p) + \varepsilon \quad (38)$$

The f_j functions, as is the case in the linear regression model with the coefficients of regression, describe the effects of each independent variable, and it is important to detect whether their inclusion in the model significantly improves it or not.

Once the model has been fitted, the additivity property of the effects enables us to examine and to evaluate separately the particular way in which each variable affects the response.

We have seen, in the two-dimensional case (X, Y), how to find a smoothed function that adapts to the trend or trace of a set of two-dimensional data (x_i, y_i). Forms of non-parametric regression such as smoothing splines can now be considered candidates to represent, in a simultaneous way, the effects produced on a variable Y in a model with multiple independent variables X1, …, Xp.

The problem now encountered is that of finding the smoothing parameters simultaneously. The most commonly adopted method consists of estimating each term using a smoothing parameter.

An iterative solution has been proposed to fit these f_j functions, namely the backfitting algorithm. In general terms, the reasoning underlying this is based on carrying out two-dimensional fits, such as smoothing splines or local regressions, to two-dimensional data that are successively generated. If we assume, in principle, that the model is correct, that is, if

$Y = \alpha + f_1(X_1) + ... + f_p(X_p) + \varepsilon$ and the corresponding f_j terms (j=1, ..., p) are optimum, then it is acceptable to assume that the expectation of the residuals derived from subtracting the sum of all the terms except the j-th one from the response will be equal to f_j.

$$E\{R_j\} = E\{\ Y - [\alpha + f_1(X_1) + ... f_{j-1}(X_j) + f_{j+1}(X_{j+1}) + ... + f_p(X_p)]\} = f_j(X_j) \quad (39)$$

Therefore, (X_j, R_j) would be well represented by a non-parametric regression curve of the type described. In practice, we begin with an initial solution (a non-parametric curve for each term) and then iteratively obtain new estimations for each f_j, fitting non-parametric curves, $\hat{f}$, to the partial residuals $\hat{R}_j$, which are updated at each step, and eliminating all the effects of the other variables from Y before performing the smoothed fit

$$\hat{f}_j(X_j) = \hat{f}(\hat{R}_j) = \hat{f}\left(Y - [\hat{\alpha} + \hat{f}_1(X_1) + ... \hat{f}_{j-1}(X_j) + \hat{f}_{j+1}(X_{j+1}) + ... + \hat{f}_p(X_p)]\right) \quad (40)$$

where $\hat{f}_j(X_j) = \hat{f}(\hat{R}_j)$ is a smoothed non-parametric regression curve for the response $\hat{R}_j$ on the independent variable X_j. The process ends when the solution stabilises.

Generalised additive models

In a similar way to the extension of the linear model to additive models, we can consider that of the generalised linear model (GLM) to the generalised additive model (GAM), assuming instead of the systematic component or linear predictor $\eta = \alpha + \beta_1 X_1 + ... + \beta_p X_p$, with the link function $g(\mu) = X\beta$, a non-linear component of the form $\alpha + f_1(X_1) + ... + f_p(X_p)$.

Broadly speaking, the fitting of a generalised additive model is based upon fitting a GLM by means of an iterative process of weighted least squares, substituting the steps concerning the weighted fits of parametric linear regression with steps of non-parametric additive regression, after having modified the algorithm to fit a weighted additive model.

9. Deriving the density function by fitting a generalised additive model (GAM)

Some authors have proposed the strategy of estimating the density function of a variable X by means of regression analysis. In this context, the independent variable is comprised of the points in the range of values observed in X that represent the mean points of the rectangles making up the histogram that describes the data sample, with intervals of equal amplitude, a. The dependent variable is formed by the corresponding heights of the rectangles, obtained as the ratios between the numbers of observations in each interval and the corresponding amplitude of the latter.

Given the sample size, n, it can be assumed that the number of observations lying within the i-th interval follows a binomial model $B(n,p_i)$, where p_i is equal to the ratio between the number of observations in the interval and n. For a large n and a small p, the binomial model is capable of approximating a Poisson model, and so we may assume that for the centre of the i-th interval, x_i, we have the value of the variable $Y=n_i$ = the number of observations in the interval i.

Therefore, a generalised additive regression model can be applied to the set of data (x_i, n_i).

The generalised linear model

$$\log(n_i) = \beta_0 + \beta_1 x_i \tag{41}$$

is not flexible enough to fit the density curve. If, instead of the linear expression as a function of x_i we take the curve $s(x_i)$, such that

$$\log(n_i) = s(x_i) \tag{42}$$

the resulting fitted curve enables us to obtain the frequencies estimated for each interval, from which we can derive the corresponding heights of the rectangles of the histograms, by dividing by the product of the sample size and the amplitude of the interval.

$$\hat{f}(x_i) = \frac{e^{s(xi)}}{n \cdot a} \tag{43}$$

where a= amplitude of the interval.

To achieve acceptable results, the sample size and the number of intervals must be large. Although the procedure is of most interest for statistics where it is more difficult to identify the shape of the density function, especially in cases where the curve may present various modes and perhaps biased behaviour patterns, we shall apply it here to show, for example, the distribution of the data resulting from a bootstrap simulation of the sampling means of the H-P residuals recorded at 64 years of age.

In total, there were 9999 values of the means of the H-P residuals, from which we obtained a frequency table of 100 intervals of equal amplitude, of approximately a=0.00001.

The following figure shows the results obtained.

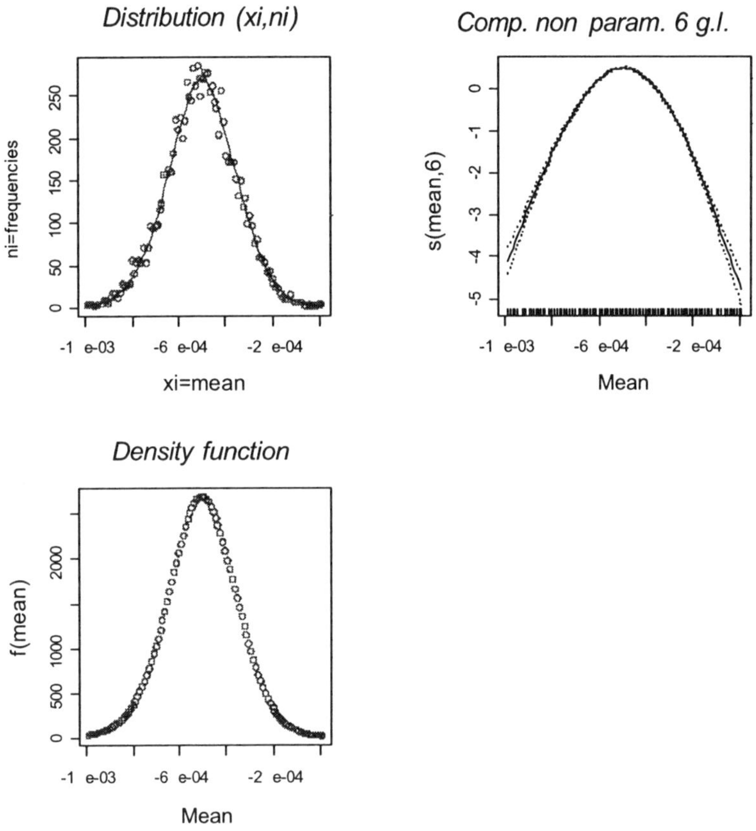

Figure 15. Result for a bootstrap simulations of the sampling means of H-P residuals (64 years)

In the next figure, we compare the density obtained by the above-described procedure with the density function corresponding to a normal distribution in which the mean and the variance coincide with those of the data.

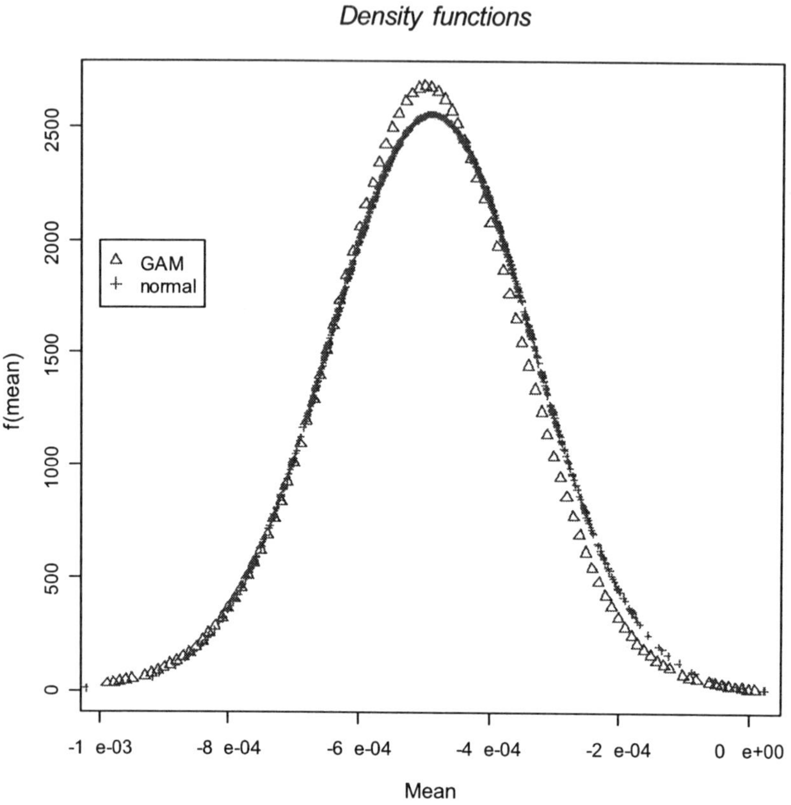

Figure 16. Result for generalized additive model and normal

10. Approximation of the distribution using saddlepoint

In practice, various methods have been applied to approximate the distribution of a statistical measure. One of the most commonly used of these is the normal approximation, although this does not always provide accurate results. A useful tool for describing distributions, and which is sometimes utilised, is the so-called saddlepoint technique, which generally provides good approximations even with small samples and in the tails of distributions, being based on the generator function of cumulants.

Although this technique was first used in the 1930s, it has recently become popular again as a means of approximating the density function. It is, in fact, a refinement of the Edgeworth expansion, which is frequently used to approximate an unknown distribution for which the moments are known. This technique gives good results in the centre of the distribution but sometimes leaves much to be desired in the tails, and can even give negative results for the density in such zones.

The derivation of the density function and the distribution function is based on the cumulant generator function K(t) and on its first two derivates with respect to t, K'(t) and K''(t).Therefore, it requires the cumulant generator function to have a known, manageable shape, a fact that means it cannot be widely used in practice. Moreover, it is necessary to numerically resolve the so-called saddlepoint equation for each value of the variable of interest.

The cumulant generator function K(t) of a variable X is given by the logarithm of the moment generator function m(t).

$$m(t) = E(e^{tX}) = \int e^{tx} f(x)dx \tag{44}$$

$$K(t) = \log\left\{E(e^{tX})\right\} \tag{45}$$

The general procedure for a saddlepoint approximation to the density and distribution functions of a statistic Y, expressed as a linear combination $Y = \sum_i a_i X_i$ of n random variables X_1, X_2, …, X_n which are identically and independently distributed with the F distribution, is as follows:

Let $K_X(t)$ be the cumulant generator function for each variable X_i from which it is possible to obtain that corresponding to Y, using $K(t) = \sum K_X(ta_i)$.

For every value of the variable Y, y, for which we wish to approximate the distribution function $F_Y(y)$, and the density function, $f_Y(y)$, it is necessary to resolve the saddlepoint equation K'(t)=y, the solution to which $t=t_y$ can be obtained, for example, by Newton-Rapson.

Different forms of the saddlepoint method are used in practice, one of the simplest being the Lugannani-Rice and Barndorff-Nielsen methods of approximating the distribution function. These are given, respectively, by the following formulas:

$$P(Y < y) = F_{sadd}(y) = \Phi(w) + \phi(w)\left[\frac{1}{w} - \frac{1}{v}\right] \tag{46}$$

$$P(Y < y) = F_{sadd}(y) = \Phi\left[w + \frac{1}{w} \log\left[\frac{v}{w} \right] \right] \tag{47}$$

where the functions Φ and ϕ are, respectively, the distribution function and the density of the standardised normal, and the values w and v are obtained from the value of the solution of the saddlepoint equation, t_y, from

$$w = sign(t_y)\left[2\{t_y y - K(t_y)\}\right]^{1/2} \tag{48}$$

$$v = t_y\left[K''(t_y)\right]^{1/2} \tag{49}$$

The saddlepoint density function is approximated using:

$$f_{sadd}(y) = \left[2\pi K''(t_y)\right]^{1/2} e^{K(t_y) - t_y y} \tag{50}$$

In particular, in the context of the replacement of a sample $X_1, X_2, \ldots, X_n$ where X_i, is selected with a probability $p_i = 1/n$, we can assume a multinomial distribution with a sum equal to n, given by the variables $(n^*_1, n^*_2, \ldots, n^*_n)$ that describe the number of times that $(X_1, X_2, \ldots, X_n)$ appears, and the mean sample statistic given by the linear combination

$$Y^* = \sum_i \frac{n_i^* X_i}{n} = \overline{X} \tag{51}$$

with $a_i = x_i/n$ which has a cumulant generator function given by

$$K(t) = n \log\left[\sum p_i e^{\frac{tx_i}{n}} \right] \tag{52}$$

The corresponding saddlepoint equation for a point x_0 in the range of $\overline{X}$ is given by

$$K'(t) = \frac{\sum e^{\frac{tx_i}{n}} x_i}{\sum e^{\frac{tx_i}{n}}} = x_0 \tag{53}$$

Of more interest is the application of these results to the bootstrap techniques with the linear approximation of a statistic, using the values of empirical influence. For example, if we approximate the T* statistic by

$$T^* = t + \sum_{i=1}^{n} \frac{n_i^* l_i}{n} \tag{54}$$

then T*-t can be expressed as the linear combination of the n^*_i, with $a_i = l_i/n$, where l_i are the influence values of the statistic.

The following figure shows the density of the variance statistic for the H-P residuals corresponding to the age of 75 years. It can be seen that the normal density does not produce such a good fit as does the saddlepoint approximation, especially in the tails.

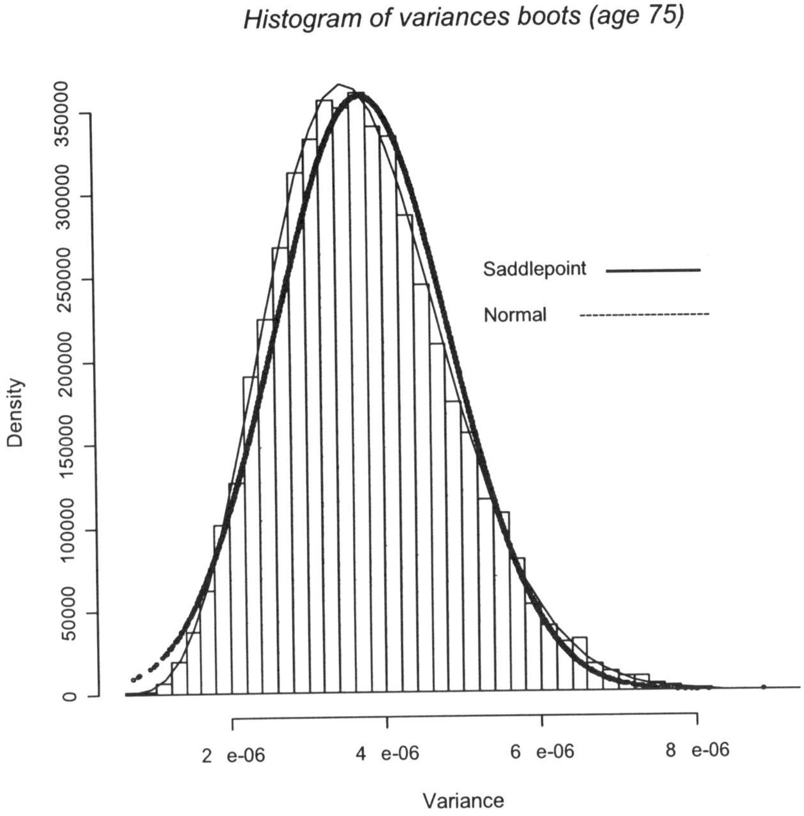

Figure 17. Distributions saddlepoint and normal

11. Conclusions

The most important source of heterogeneity in the residuals is not in the sets generated by the different curves that are fitted for each year, but within each curve, in those generated for different ages.

The mixing of residuals into a single set reveals a distribution with a behaviour pattern far removed from the normal one. The different means and dispersions, corresponding to residuals derived from fits for different ages, give rise to a distribution with various modes.

The following figures show the results of distributions of the simulated mean and variance estimators of the total set of H-P residuals, without distinguishing by age or period.

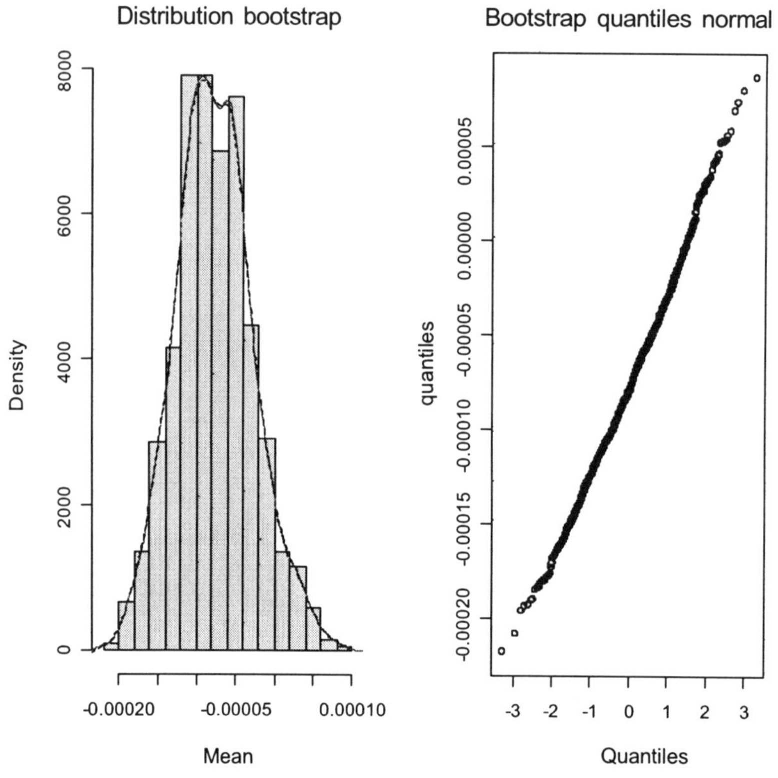

Figure 18. Distributions of simulated means variances

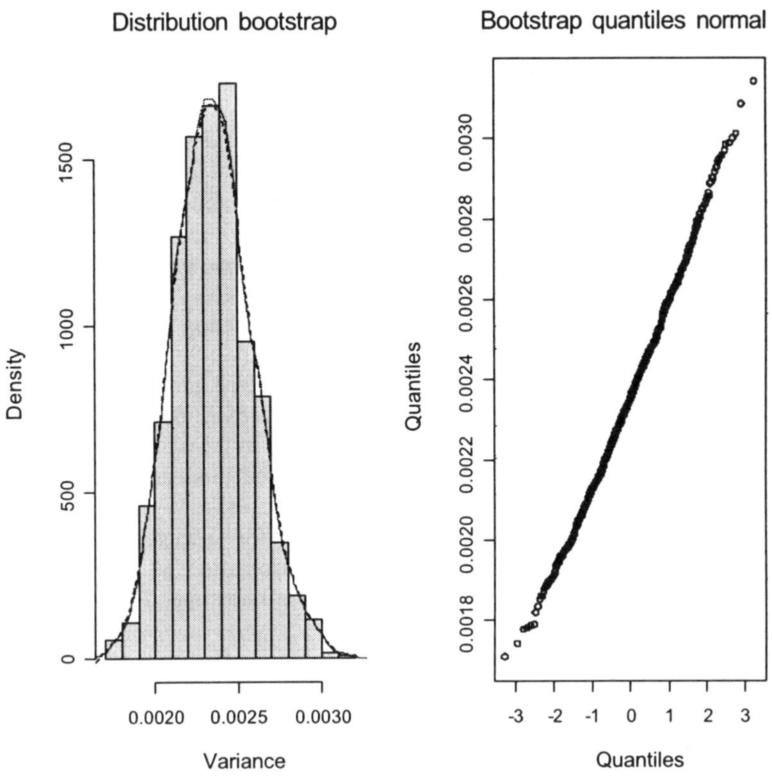

Figure 19. Distributions of simulated variances

Exploration of the distributions of certain statistical measures of interest enables us to evaluate behaviour patterns. Graphic techniques, as well as those for fits for models of greater or lesser complexity, and particularly non-parametric techniques, can be complementary and constitute useful tools for performing this task, in which we seek to discover how schemas for modelled structures (the Heligman and Pollard curve) are adapted to reality (the mortality rates observed).

References

1. Booth, J.G., Hall, P. and Wood, A.T.A. (1993). Balanced importance resampling for the bootstrap. Annals of Statistics, 21, 286–298.

2. Davison, A.C., Hinkley, D.V. and Schechtman, E. (1986). Efficient bootstrap simulation. Biometrika, 73, 555–566.
3. Davison, A.C. and Wang, S. (2002). Saddlepoint approximations as smoothers. Biometrika, 89(4), 933–938.
4. DiNardo, J. and Tobias, J.L. (2001). Nonparametric density and regression estimation. Journal of Economic Perspectives, 15(4), 11–28.
5. Efron, B. (1990). More efficient bootstrap computations. Journal of the American Statistical Association, 55, 79–89.
6. Efron, B. (1986). How biased is the apparent error rate of a prediction rule? Journal of the American Statistical Association, 81, 461–470.
7. Efron, B. and Tibshirani, R. (1993). An Introduction to the Bootstrap. Chapman & Hall.
8. Efron, B. (1992). Jackknife-after-bootstrap standard errors and influence functions (with Discussion). Journal of the Royal Statistical Society, Series B, 54, 83–127.
9. Gleason, J.R. (1988). Algorithms for balanced bootstrap simulations. American Statistician, 42, 263–266.
10. Johns, M.V. (1988). Importance sampling for bootstrap confidence intervals. Journal of the American Statistical Association, 83, 709–714.
11. Hall, P. (1989). Antithetic resampling for the bootstrap. Biometrika, 73, 713–724.
12. Hinkley, D.V. (1988). Bootstrap methods (with Discussion). Journal of the Royal Statistical Society, Series B, 50, 312–337, 355–370.
13. Hinkley, D.V. and Shi, S. (1989). Importance sampling and the nested bootstrap. Biometrika, 76, 435–446.
14. Kuonen, D. (1999). Saddlepoint approximations for distributions of quadratic forms in normal variables. Biometrika.
15. McCullagh, P. and Nelder, J.A. (1989). Generalized linear models. Chapman & Hall.
16. Rust, R.T. (1988). Flexible Regression. Journal of Marketing Research.
17. Sheather, S.J. and Jones, M.C. (1991). A reliable data-based bandwidth selection method for kernel density estimation. Journal of the Royal Statistical Society, Series B, 53, 683–690.
18. Silverman, B.W. (1985). Some aspects of the spline smoothing approach to non-parametric curve fitting. Journal of the Royal Statistical Society, Series B, 47, 1–52.
19. Stone, M. (1974). Cross-validation choice and assessment of statistical predictions (with Discussion). Journal of the Royal Statistical Society, Series B, 36, 111–147.
20. Terrell, G.R. (1998). The Gradient Statistic. Statatistic Department, Virginia Polytechnic Institute and State University. Blacksburg, Virginia.
21. Terrell, G.R. (2003). A stabilized Lugannani-Rice formula. Department of Statistics, BPI&SU, Blacksburg, Virginia.

22. Wang, S. (1995). One-step saddlepoint approximations for quantiles. Computational Statistic and Data Analysis.
23. Wolf, C.A. and Summer, D.A. (2001). Are farm size distributions bimodal? Evidence from kernel density estimates of dairy farm size distributions. American Agricultural Economics Association.
24. Wu, J. and Wong, A.C.M. (2003). A note on determining the p-value of Bartlett's test of homogeneity of variances.

Chapter 15

MEASURING THE EFFICIENCY OF THE SPANISH BANKING SECTOR: SUPER-EFICIENCY AND PROFITABILITY

J. GÓMEZ-GARCÍA

Department of Quantitative Methods for Economics, University of Murcia
Campus de Espinardo, s/n, Espinardo 30100 Murcia, Spain

J. SOLANA-IBÁÑEZ

Department of Business and Management, Catholic University San Antonio of Murcia
Campus de los Jerónimos, s/n, Guadalupe 30107 Murcia, Spain

J.C. GÓMEZ GALLEGO

Department of Business and Management, Catholic University San Antonio of Murcia
Campus de los Jerónimos, s/n, Guadalupe 30107 Murcia, Spain

We analyse the dependent relationship between technical efficiency and profitability of commercial banks in Spain, using multivariate techniques, such as factorial, cluster and discrimination analyses. Efficiency measurements are obtained by Data Envelopment Analysis (DEA), incorporating size and management-related variables as inputs and outputs. Efficiency and super-efficiency coefficients are obtained for each bank and conclusions are made concerning the existence of differing levels of profitability according to the efficiency level with which the banks are managed.

1. Introduction

The high degree of correlation between the behaviour of the economy and the banking sector, together with the sector's role as financial intermediary, Pastor [1], is ample reason for the continual interest in different aspects of the banking system.

Traditionally, this kind of study has been approached through the use of costs and profitability ratios, Pastor, Pérez and Quesada [2], although more recently these traditional techniques have tended to be replaced by the use of econometric techniques that look at an institution from a global viewpoint that considers the inputs used and outputs obtained, that is techniques that permit the efficiency of an organization to be measured. One such technique is that known as Data Envelopment Analysis (DEA), a non-parametric econometric technique that permits the way in which a company is managed to be evaluated more thoroughly.

From the end of the 1980's, there has been a growing interest in the importance of X-type* efficiency as opposed to scale efficiency in the banking sector. The fact that several studies of the banking sector have shown that the spread of mean costs was greater among banks of a given size than among banks of different sizes, points to the greater importance of reducing X-type inefficiencies rather than of attaining an optimal production size (economies of scale) as a means of reducing costs. After this analysis of the relation between efficiency and costs, interest turned to the analysis of the relation between efficiency and the profitability of different banks.

Economics theory says that companies wishing to maximize profits must produce at the minimum cost possible. In other words, obtaining the maximum level of profits and, hence, attaining maximum profitability involves being economically efficient. Berger [3] in the USA, Goldberg and Rai [4] in Europe and Maudos and Pastor [5] in Spain, suggested that efficient banks are generally more profitable than inefficient banks.

Berger and De Young [6] demonstrated that efficiency, as indicator of management quality, influences the assignation of lendable funds between clients. Since, according to Freixas [7], the rate with which clients fall into arrears contributes to explaining the evolution of profitability in the banking sector, the effect of efficiency on profitability not only influences the reduction of costs but also has implications on the process of giving credit. Efficiency, then, not only has an effect on profitability derived from the reduction of costs, but also derived from the management quality that any efficient bank enjoys and that manifests itself in many banking spheres.

This study represents an analysis of the efficiency and profitability of Spanish banks. In so far as the measurements of efficiency is based on DEA, it is closely interrelated with a research line looking at methods of analysing global efficiency and the establishment of a ranking of efficiency. The study focuses on the relation between the productive efficiency and profitability of banks. For this reason, in the second part we describe the methodology of estimating the productive efficiency of banks. In the third section, we shall describe the way in_which banks were sampled and the statistical sources used. In the fourth section, we shall relate the profitability of a company with its super-efficiency and, lastly, we shall present our conclusions.

---

*So called X-type inefficiencies are those due to errors in management and/or organization, and include technical inefficiencies such as the allocative type, and differ from scale inefficiencies.

2. Reach of the study

The data base used is that published by the Spanish Banking Association (AEB) [8], in Spanish for 2002 and 2003, which provided ample information on the different characteristics concerning the type and volume of activity of Spanish banks. However, the wide-ranging specializations of many banks meant that certain areas were not covered and led us to choose a group of 36 for which complete information on the selected variables was available, table 1.

For each variable we estimated minimum, mean, typical deviation and asymmetric coefficient. Of note is the different degree of representativeness of the mean value of each variable since, as can be seen, the typical deviations took on extreme values, which were very small or very large. The non-ratio variables were widely dispersed and the asymmetric coefficient pointed to generally very asymmetric distributions, except in the case of CROF, ICAT and ROE; in the case of DPRA, the asymmetry was significantly negative.

Table 1. Descriptive statistics

Variables	Code	Unit	Min	Mean	S.D.	As.
Mean T. Assets.	ATM	E. 10^3	104928.0	16383920.3	44228723.4	3.8
Cashiers	CAJ	Unit.	.0	536.6	1086.8	2.9
Current Acc.	CC	Unit.	826.0	268718.5	504860.0	2.6
Credits	CRD	E. 10^3	54540.0	10101139.3	22688793.9	3.6
Debts	DEB	E. 10^3	7899.0	9424198.0	21553846.8	3.5
Employees	EMP	Unit.	26.0	2922.5	6294.74	3.5
Expl. Margin	MEX	E. 10^3	-1373.0	434417.9	1235077.7	3.7
Int. Margin	MIN	E. 10^3	2227.0	614255.7	1702052.5	3.8
Net	NET	E. 10^3	22841.0	1099687.3	3020618.6	4.4
Cards	TAR	Unit.	.0	622977.3	1429731.1	3.7
Cashiers/Offic	CAOF	Unit.	.00	.89	.65	.38
Créd./Offic	CROF	E. 10^3	-52.63	11.60	43.26	3.51

Table 1. (*Continued*)

Variables	Code	Unit	Min	Mean	S.D.	As.
Dep./Offic	DPOF	E. 10^3	343.43	116919.07	442366.0	5.61
Dep./R.Aj	DPRA	%	.51	.91	.08	-4.07
Empl./Offic	EMOF	Unit.	1.18	13.26	24.61	4.37
Cred./ATM	ICAT	%	.09	.60	.26	-.45
Cred./Emp	ICEM	E. 10^3	371.02	10378.3	34920.2	5.23
Expl M./ATM	MEATM	%	-1.31	2.08	1.97	1.53
Int. M./ATM	MIATM	%	.32	3.41	2.35	1.59
Net /ATM	NEATM	%	.02	.08	.08	4.47
ROE	ROE	%	-.08	.17	.15	1.1
Card./SharHdr.	TARAC	Unit.	.00	16777.2	68808.8	5.42
Card/Offic	TAROF	Unit	.00	51456.01	270029.8	5.86

3. Methodology

3.1. *DEA Origin and diffusion*

Economics and Operational Research share many interests, one of the most important being analysis of the production possibilities of a productive unit. The definitive connection arose in 1978 from the work of Abraham Charnes, William W. Cooper and Edward Rhodes [9] (CCR) entitled "*Measuring the Efficiency of Decision Making Units*", published in the "European Journal of Operations Research". The DEA model that they presented led to growing popularity of the empirical use of lineal programming techniques for calculating coefficients of efficiency; so much so that by 1999 the work of CCR had been cited more than 700 times in SSCI.

The starting point was the seminal work of Michael James Farrell [10], "*The Measurement of Productive Efficiency*", published in the "Journal of the Royal Society" in 1957, where the concept of efficiency was first mooted.

The most influential work related with such aspects of macroeconomics was that of Solow [11] published in the "Review of Economics and Statistics" and entitled "*Technical change and the aggregate production function*". At the same time Farrell established the bases for studying efficiency and productivity at microeconomic scale, putting forward two novel aspects: how to define efficiency and productivity, and how to measure efficiency.

Faced with the possibility of inefficiency, Farrell opted for the concept of border_production as opposed to the mean efficiency underlying most of the econometric literature to date on the production function. The new focus of Farrell consisted of decomposing efficiency into technical and allocative efficiency at individual production unit level. The radial contraction/expansion connecting inefficient units with efficient units with respect to the production function constitutes the base for measuring efficiency and is the true contribution of Farrell.

Farrell proposed a measure of efficiency consisting of two components: technical efficiency and assignative efficiency, both of which combine to provide a measure of total economic efficiency. These measures assume that the production function of efficient companies is known. Since this function is never known in practice, as Farrell recognized, he proposed two possibilities: obtaining a non-parametric function or a parametric function.

The first alternative gave rise to the models of estimating non-parametric frontiers and was followed by Charnes, Cooper and Rhodes [9], and resulted in an approach to DEA. A subsequent model gave rise to a great quantity of research and was denominated FDH (free disposal hull) formulated in 1984 by Deprins, Simar and Tulkens [12], and developed by Tulkens [13] in 1994. This second pathway was followed by Afriat [14] and Aigner [15], resulting in two approximations known as the determinist and stochastic frontier models.

An intermediate pathway comprising models that we might term models that do not use production frontiers, is provided by index numbers, and their use in measuring efficiency and productivity is indirect. They are used, rather, to generate variables or data that can be used in the application of the DEA models or in the estimation of stochastic frontiers, Solana [16].

3.2. *Models*

Since its genesis, Charnes *et al* [9] have developed a variety of DEA models, both input and output oriented, depending on the existence of constant or variable returns (in this last case, depending, too, on whether these are growing or diminishing) and whether the inputs can or cannot be controlled, among other aspects. The first model we applied was that initially proposed by Charnes *et al* [9] and known as CCR, after its authors. This model implies returns on a constant scale and is input oriented. In accordance with Cooper *et al* [17], the starting point is the traditional definition of efficiency (coefficient between outputs and inputs) and the aim is, by means of lineal programming, to obtain weights so that, the ratio between outputs and inputs can be maximized.

To calculate the efficiency of n units, n lineal programming problems must be solved to obtain both the values of the weights (v_i) associated with the inputs (x_i), and the weights (u_r) associated with the outputs (y_r). Assuming m inputs and s outputs, and transforming the fractional programming model into a lineal programming problem, the input oriented CCR model is formulated as follows:

$$Max \quad \theta = u_1 y_{1o} + u_2 y_{2o} + \ldots + u_s y_{so}$$

s.a.

$$v_1 x_{1o} + v_2 x_{2o} + \ldots + v_m x_{mo} = 1 \qquad (1)$$

$$u_1 y_{1j} + u_2 y_{2j} + \ldots + u_s y_{sj} \leq v_1 x_{1j} + v_2 x_{2j} + \ldots + v_m x_{mj} \quad j = 1,2,\ldots,n$$

$$v_i \geq 0 \quad (i = 1,2,\ldots,m)$$

$$u_r \geq 0 \quad (r = 1,2,\ldots,s$$

The output oriented lineal version is formulated as follows:

$$\text{Min} \quad \rho = v_1 x_{1o} + v_2 x_{2o} + \ldots + v_m x_{mo}$$

s.a.

$$u_1 y_{1o} + u_2 y_{2o} + \ldots + u_s y_{so} = 1 \qquad (2)$$

$$u_1 y_{1j} + u_2 y_{2j} + \ldots + u_s y_{sj} \leq v_1 x_{1j} + v_2 x_{2j} + \ldots + v_m x_{mj} \quad j = 1,2,\ldots,n$$

$$v_i \geq 0 \quad (i = 1,2,\ldots,m)$$

$$u_r \geq 0 \quad (r = 1,2,\ldots,s)$$

Given the lack of information on the form of the production frontier, we have used models analogous to those in [1] and [2] but which permit variable returns, and known as BCC, after its authors, Banker *et al* [18]. In this work, we use the output oriented model (BCC-O), formulated as:

$$\text{Min} \quad Z = \sum_i v_i x_{io} + v_o$$

s.a.

$$\sum_r u_j y_{ro} = 1$$

$$\sum_i v_i x_{ij} - \sum_r u_r y_{rj} + v_o \geq 0 \quad j = 1,2,...,n \tag{3}$$

$$v_i \geq 0 \quad (i = 1,2,...,m)$$

$$u_r \geq 0 \quad (r = 1,2,...,s)$$

$$v_o \quad \text{free}$$

where v_o is the variable that permits us to identify the nature of the scale returns to scale. To obtain a more complete ranking, efficient units are classified by applying the MDEA models proposed by Andersen, P. and Petersen, N. C. [19].

3.3. *Efficiency of banking management*

As explained by Thanassoulis [20], banking institutions have two activities whose efficiency can be analysed: production and intermediation.

The efficiency of production refers to how banks are used: labour, capital, space, service accounts, etc., all of which is reflected in a wide range of transactions such as the search for resources, the process of advancing credit and other income-generating activities.

The choice of inputs and outputs is a controversial subject that presents several problems, since the products of banks are immaterial, heterogeneous and jointly produced. Furthermore, this heterogeneity is continually changing: not only do new products appear and disappear, but the proportions of the components of the output vector also change.

Two basic solutions have been proposed to resolve this problem, Pastor [21]:

- The first involves measuring output by adding given sections of the balance sheet of the institutions (deposits, total assets, loans, etc.). This is known as a monetary focus, according to which the total assets and/or deposits are magnitudes representative of financial services and payments, respectively. This approach has the advantage of simplicity and the availability of relevant data, so that it is frequently used in studies of economics of scale.
- The second solution, known as the physical or non-monetary focus, equates banking activity with the productive processes of industrial companies by using magnitudes, such as the number of loans and deposits, etc., which are the equivalent of the number of service units offered. This approach is very suitable for studying some aspects associated with cost-size relations. The

lack of information makes it difficult to apply, at least in the case of Spanish banks.

In this work we consider a bank as a company producing a flow of services, which involves the consumption of inputs. This flow of services, associated with active and passive items, will constitute the measurement of ideal output.

The choice of volume of credits and loans as basic representative measures of input and output supposes that these factors provide the clients of assets and liabilities with greater fluidity of resources and services. The conceptualization of a bank as a company that produces services, and the use of proxy variables, such as deposits and loans, normally associated to the provision of such services, obliges us to consider an additional output that is closely related with the conditions of providing these services: the number of current accounts.

Taking into consideration the cited literature, and the information available in the case under study, the variable selected as inputs and outputs are the following: Current Accounts, Intermediation Margin, Net Profit, Debits.

4. Results

When a wide number of correlated variables are available for a given population, factorial analysis (FA) permits the information contained in these variable to be synthesized into a lower number of variables (factors).

After typifying the original variables and demonstrating the existence of a significant correlation, Barlett's sphericity test and the statistics of Kaiser-Meyer-Olkin (KMO) were applied. These gave a Chi-squared value of 1283.59, with 120 g.l., for the Bartlett test and a KMO value for the sample suitability of 0.637, with an associated significance level of 0.000. Next, the factorial axes were extracted by principal components analysis. Lastly, the axes chosen were rotated by Varimax to facilitate understanding.

Of the original variables observed, those related with size, profitability, management and risk were selected, Moya and Caballer [22]. In this way the following fifteen variables were included: Current Accounts, Debits, Time deposits, ATM, Intermediation Margin, Intermediation Margin on ATM, ROE, Operating Margin, Operating Margin over ATM, Credit investment over employee, Credit Investment over ATM, Net profit over AMT, Debits per employee, Deposits over debt capital.

Applying the FA procedure, four factorial axes were obtained which explained 91.27% of the global variance. These were chosen bearing in mind the value of the autovalues of the characteristic equation, in accordance with the criterion of the arithmetic mean. From the matrix of rotated components, the factorial axes were defined as follows:

Factor 1: saturated by C/C (.900), CRD (.989), DEB (.995), IMPL (.984) ATM (.986), ME (.965), MI (.967). Factor 2: saturated by MEATM (0.983), MIATM (.942), ROE (.623), BATM (.932). Factor 3: saturated by ICEMP (.970), DBEMP (.986). Factor 4: saturated by ICATM (.729), DPRA (.820).

Factor 1, with an associated autovalue of 7.09, explains 47.26% of the total variance; Factor 2, with an associated autovalue of 3.21, explains 21.41% of the total variance; Factor 3, with an associated autovalue of 1.37, explains 13.28% of the total variance, and Factor 4, with a characteristic root of 1.37, explains 9.18% of the total variance.

From the correlations between the factorial axes and the original variables, we have interpreted and, consequently, denominated the factorial axes as follows: Factor 1: Size; Factor 2: Profitability; Factor 3: Management And Factor 4: Risk.

Applying the BCC-O model, efficiency coefficients were obtained which situated 14 financial institutions in the frontier, while the remaining 22 had a different percentage of technical inefficiencies. The MDEA-O model was used to establish a complete ranking, and the corresponding coefficients of super-efficiency were obtained for each bank.

4.1. *Analysis of the relation efficiency-profitability*

Appling the Cluster Analysis procedure to the efficiency and super-efficiency distributions, the banks were grouped into homogeneous conglomerates to study the profitability of these groups. In this work, we apply the hierarchical grouping method, whereby the groups themselves are paired, so that there is a lower increase in the total distances. The different cluster levels are established by taking into account the value of intra-group variance. Three highly homogeneous groups were formed since the coefficient of variability did not exceed 20%. To test the suitability of the grouping obtained in the cluster analysis, we applied discrimination analysis, obtaining the discriminant function from the values of efficiency and super-efficiency. The results confirmed that classification was correct in 100% of cases.

Table 2 contains information on the financial institutions of each cluster, the mean values and typical deviations obtained for the variables, profitability and super-efficiency. Also contains the coefficients of efficiency for each bank, and super efficiency if necessary.

Table 2. Cluster, mean values, typical deviations, coefficients of efficiency-super efficiency

Clusters	Efficiency	Profitability	Banks (Efficiency) - (*Super-efficiency)	
			Sabad. BPr. (500)*	Patagón (451.8)*
1	Mean: 469.96	Mean: -0.85	Popular BP. (500)*	Créd.Local.(500)*
N=5	D.T.: 45.33	D.T.: 0.61	Pueyo (398.9)*	
			Cooperat.E. (47.3)	De Pyme (36.9)
			Simeón (56.1)	Urquijo (58.4)
			Halifax (20.8)	Espirito S.(55.4)
			Barclays (78.2)	Bankoa (38.9)
			Bancofar (11.29)	Deutsche (81.2)
2	Mean: 56.10	Mean: 0.03	Gallego (55.7)	Sabadell (80.6)
N=22	D.T.: 19.89	D.T.: 0.94	Pastor (71.7)	Guipuzc. (82.3)
			Citybank (29.1)	Atlantico (83.8)
			Vasconia (55.9)	March (58.2)
			Castilla (50.6)	C.Balear (64.8)
			Galicia (53.7)	Andalucía (62.5)
			Fibanc (174.3)*	Bankinter(171.0)*
			BBVA (191.2)*	E.Crédito(220.2)*
3	Mean: 205.05	Mean: 0.40	BSCH (154.1)*	Valencia (204.0)*
N=9	D.T.: 58.70	D.T.: 1.10	Popular Esp (134)*	Banif (305.2)*
			Santan.C.F.(290)*	

ANOVA was applied to ascertain whether the three groups defined by super-efficiency differ significantly as regards their levels of profitability. Significantly different (p=0.043) levels of profitability were found. Applying Bonferroni's test, differences in profitability were seen between groups 1 and 3 (p=0.038), while a p-value of 0.056 was observed for groups 1 and 2, and a p-value of 0.356 for groups 2 and 3.

5. Conclusions

Factorial analysis provided four factors that encapsulated all the characteristics of Spanish banks; size, management, profitability and risk. In this way, each bank is represented in a tetra-dimensional space by the vector which components are the scores of the bank on each of the four factorial axes.

Applying DEA analysis, the BCC-O model and MDEA, we obtained an efficiency ranking for 36 financial institutions. In 22 banks we observed a percentage of technical inefficiency. Changes in the way of managing these banks could bring them to the production frontier.

Applying cluster analysis to the super-efficiency scores enabled us to make homogeneous groups of the banks analysed. In this way we obtained three groups of minimal intra-group variance.

As regards profitability, we conclude that there are significant differences between the groups of banks established from the measures of super-efficiency. Significantly different (p=0.043) levels of profitability were found. Group 3, with high level of medium efficiency (205.05), presents the highest level of profitability, while, group 1, (469.96), shows quite low profitability. Despite the significant differences, we need to take into account other characteristics such as specialization, in order to explain these findings. This could be the topic of future research.

References

1. J.M. Pastor. (1998). Gestión del Riesgo y Eficiencia en los Bancos y Cajas de Ahorros, Serie Documentos de Trabajo, No 142/1998. Fundación de Cajas de Ahorro Confederadas para la Investigación Económica y Social España.
2. J.M. Pastor, F. Pérez and J. Quesada. (1995). Are European Banks Equally Efficient? Revue de la Banque, June, 324–33.
3. A.N Berger. (1995). The Profit-Relationship in Banking - Tests of Market-Power and Efficient-Structure Hypotheses. Journal of Money, Credit and Banking, 27(2), 405–431.
4. L.G. Goldberg and A. Rai. (1996). The structure-performance relationship for European banking. Journal of Banking and Finance, 20, 745–771.
5. J. Maudos and J.M. Pastor. (1998). La eficiencia del sistema bancario español en el contexto de la Unión Europea. Papeles de Economía Española, 84/85, 155–168.
6. N. Berger and R. De Young. (1997). Problem Loans and Cost Efficiency in Commercial Banks. Journal of Banking & Finance, 21(6), 849–870.

7. X. Freixas, J. De Hevia and A. Inurrieta. (1993). Componentes macroeconómicos de la morosidad bancaria: un modelo empírico para el caso español. Moneda y crédito, 99, 125–156.
8. Anuario Estadístico de la Banca en España. (2003). Asociación española de banca.
9. A. Charnes, W.W. Cooper and E. Rhodes. (1978). Measuring the efficiency of decision-making units. European Journal of Operational Research, 2, 429–444.
10. M. J. Farrell. (1957). The measurement of productive efficiency. Journal of the Royal Statistical Society, Series A, 120(III), 253–281.
11. R.A. Solow. (1957). Technical chance and the aggregate production function. Review of Economics and Statistical, 39, 312–320.
12. D. Deprins, L. Simar and H. Tulkens. (1984). Measuring labour efficiency in post offices. The performance of public enterprises: Concepts and measurement, Marchand, M., Pierre Pestieau and Henry Tulkens, ed., 243–267 Amsterdam, North Holland.
13. H. Tulkens. (1994). On FDH analysis: Some methodological issues and applications to retail banking, courts and urban transit. Journal of Productivity Analysis, 4(1–2), 183–210.
14. S. Afriat. (1972). Efficiency estimation of production functions. Economic Review, 13(3), 568–598.
15. D.J. Aigner, C.A. Knox Lovell and P. Schmidt. (1997). Formulation and estimation of stochastic frontier production function models. Journal of Econometrics, 6(1), 21–37.
16. J. Solana. (2003). Modelos DEA para la evaluación global de la eficiencia técnica. Obtención de un ranking de unidades productivas. Tesis doctoral. UCAM.
17. W.W. Cooper, L.M. Seiford and K. Tone. (2000). Data envelopment analysis: A comprehensive text with models, applications, references and DEA-solver software. Boston: Kluwer.
18. R.D. Banker, A. Charnes and W.W. Cooper. (1994). Some models for estimating technical and scale inefficiencies in data envelopment analysis. Management Science, 30, 1078–1092.
19. P. Andersen and N.C. Petersen . (1993). A procedure for ranking efficient units in Data Envelopment Analysis. Management Science, 39(10), 1261–1294.
20. E. Thanassoulis. (1999). Data Envelopment Analysis and its use in banking. Interfaces, May/June 29, Ed. 3.
21. J.M. Pastor. (1998). Diferentes metodologías para el análisis de la eficiencia de los bancos y cajas de ahorro españoles. Departament de Análisi Econòmica Universitat de Valencia.
22. Moya and V. Caballer. (1994). Un modelo analógico bursátil para la valoración de cajas de ahorro. En Hernández Mogollón, R.M. (Ed.): La reconstrucción de la empresa en el nuevo orden económico, 287–297.